Fundamentals of Water Desalination

ENVIRONMENTAL SCIENCE AND TECHNOLOGY SERIES

SERIES EDITOR

J. Carrell Morris
Gordon McKay Professor of Sanitary Engineering
Harvard University
Cambridge, Massachusetts

Volume 1: Fundamentals of Water Desalination
by Everett D. Howe

Other volumes in preparation

Fundamentals of Water Desalination

EVERETT D. HOWE

Professor Emeritus
Department of Mechanical Engineering
and
Director Emeritus
Sea Water Conversion Laboratory
University of California
Berkeley, California

MARCEL DEKKER, INC. New York 1974

MARCEL DEKKER, INC.

270 Madison Avenue, New York, New York 10016

LIBRARY OF CONGRESS CATALOG CARD NUMBER: 74-76999

ISBN: 0-8247-6200-2

Current printing (last digit):
10 9 8 7 6 5 4 3 2 1

PRINTED IN THE UNITED STATES OF AMERICA

CONTENTS

PREFACE

The purpose of this book is to provide a basic understanding of the many methods of desalting saline water. It is intended primarily for chemists, engineers, and other persons who are concerned with the technical problems in the field of producing potable freshwater. The increasing importance of desalting technology is due to population growth as well as to the need for overcoming the effects of pollution, occasional droughts, and local scarcities of high-quality freshwater in areas otherwise suitable for industrial development and habitation. It is probable, at least in the near future, that desalination methods will be widely used in the quality improvement of existing brackish or slightly saline water sources rather than in the procurement of freshwater from the sea. It is hoped that this book may contribute to the solution of a wide range of desalting problems by the ever-increasing numbers of people who have become concerned with the problems of supplying freshwater.

Those readers who are familiar with the literature in the field of desalination will be interested to note the use in the text of the compound words "freshwater" and "seawater" rather than the more familiar forms "fresh water" and "sea water." This usage recognizes the developing practice in the literature as well as the recommendation of recent editions of dictionaries.

In organization, the first four chapters contain information regarding fresh- and saline-water supplies, and conclude with brief descriptions of the several existing and potential methods for desalination of saline sources. The remaining chapters outline some of the design problems pertinent to each of the particular desalination methods described in the fourth chapter. It is intended that these chapters on particular desalination methods will help the reader to identify and be made aware of the many problems of design in order to have a basic appreciation of the concerns necessary for the planning of a complete plant.

No attempt is made in this book to give details of the computer techniques used in modern design offices to optimize a given plant design. This is beyond the scope of a book of this length and would be ineffective because of the lack of the proprietary information needed for optimization purposes. Such information is available for any given design group through company sources. The algebraic relations given in this book form the basis for such computer programs, although their presentation here is for the purpose of expressing interrelations between important variables.

The writing of this book was made possible largely through my interactions with colleagues, researchers, and practitioners in the desalting field over the past twenty years. Much was learned from my administrative and research affiliation with the Seawater Conversion Laboratory of the University of California, and from personal discussions with numerous other research workers and perusal of their published reports. Also important were extensive visits to operating plants and research establishments, both within the United States and throughout the world, all of which gave insight into the solving of practical problems peculiar to the different desalting methods. I am particularly grateful to Alan D. K. Laird, presently the Director of the Seawater Conversion Laboratory, for his encouragement during the preparation of this book. The assistance of my wife, Ruth M. Howe, is gratefully acknowledged, since she not only typed the entire manuscript but also made many suggestions for simplifying and clarifying the text material.

EVERETT D. HOWE

Fundamentals of Water Desalination

Chapter 1

WATER AND DESALINATION

I. WATER QUALITY AND ITS IMPORTANCE

Desalination has become increasingly important since about 1950 because of the tendency in recent years for the world's swelling population to dwell in areas where local supplies of high quality freshwater are less than adequate. Also, the exploitation of mineral resources in some of these areas has reached levels of production such that the monetary returns are sufficient to permit people native there to seek alternatives to the meagre supplies of brackish water which they previously had been accustomed to use. Since, in many of these areas, there are saline sources of considerable extent close at hand, the use of desalination is becoming an attractive possibility. In order to place desalination in proper perspective it is desirable to consider some of the factors affecting the demand and supply of freshwater, a commodity which is all-important to every civilized community, regardless of its degree of industrialization.

The basic need for the desalination of water arises because of the essential role of water in all organic life. It is the solvent portion of the fluid in plants and animals which carries nutrients from cell to cell and assures correct functioning of the several vital organs. The heat and waste matter resulting from the metabolic processes which constitute physical life are eliminated as water vapor and aqueous solutions, respectively, thus requiring periodic replenishment of water.

This replenishment is accomplished by the passage of water through membranes, such as through the root covering of plants and the lining of the colon in man and animals. These membranes screen out dissolved materials and osmotically permit water to pass from a dilute solution to a more concentrated one. This essential characteristic of the life process necessitates that the water supply must be less concentrated than the fluids within the plant or animal.

Because of the great solvent capacity of water, there is a tendency in many areas for it to dissolve chemical materials to such an extent that it becomes unsuitable for organic life. The concentration of these dissolved chemicals determines its quality, or "adequacy." High on the list of "poor quality" is seawater with a concentration of ions greater than plant fluid or animal blood can tolerate. Only in a few instances have fish, certain sea animals, and aquatic plants evolved special provisions for replenishing water in their systems from supplies of salinity greater than the usual norm. This is accomplished through the evolution of a specialized organ containing a fluid more concentrated than normal, thus providing tolerance for greater water salinity.

The quality of water acceptable for the average human consumption has been standardized by agencies such as the United States Public Health Service (USPHS), which has adopted requirements [1] for chemical quality as listed in Table 1-1. The objective of most desalination plants is to reduce the salinity of water to something less than 500 ppm, the maximum desirable figure for potable water, as noted above. There are additional requirements for potable water set by the USPHS. These determine the limitation of content of coliform bacteria and other materials which could directly affect the health of consumers. These requirements will not be dealt with here since they have no effect on the chemical quality of water, the subject of this book.

It must be realized that not all public water supplies meet the chemical standards noted above, and that throughout the world many people are using water somewhat more saline than the maximum 1500 ppm cited. Dixey [2] has given considerable attention to this point and has surveyed conditions in many parts of the world. He reports that water with more than 4000 ppm is used in northern Mexico, 3400 ppm in Senegal, 3130 ppm is commonly acceptable in parts of Australia, and that even 6100 ppm was used by some Australian settlers for an extended period of time. The critical element in such water supplies lies in the particular materials found in solution. Dixey concluded that calcium carbonate has no harmful physiological effect. Alkaline carbonates are most injurious while alkaline sulfates are least injurious. Alkaline chlorides, including common salt, are intermediate in

TABLE 1-1

Chemical Standards for Potable Water Supplies
USPHS Requirements

Chemical constituents	Optimum concentration in parts per million (ppm)	Maximum concentration (ppm)
Lead	0	0.1
Copper	0	3.0
Zinc	0	15.0
Iron + manganese	0	0.3
Sulfate	250.0	250.0
Chloride	250.0	
Magnesium	125.0	125.0
Fluoride	0.5	1.5
Total dissolved solids[a]	500.0	1500.0

[a]The individual concentrations in the two columns do not add up to the total dissolved solids shown at the bottom of each column. This is because each entry represents the optimum or maximum allowable concentration of the specified ions. Therefore, in the optimum column, when one ion appears near its maximum value, other ions of the same electrical sign must show lesser concentrations so that the total solution will not exceed the total dissolved solids specified. The maximum column represents the maximum values of only those ions listed, but assumes the possible presence of unspecified ions.

the above list. Magnesium sulfate (epsom salts) gives a bitter taste to water, and has its usual purgative affects; iron in amounts as small as 0.5 ppm is offensive to the taste. There is no indication of the number of people using these low-grade waters and it is possible that the number may be small in some of the cases.

Surveys of the municipal water supplies in the United States were made in 1964 and 1969 [3] under the auspices of the Office of Saline Water. These were limited to growing communities serving at least

1000 people. In the tabulated replies from some 4700 cities, about 10% of those which included data on chemical content showed salinities greater than 500 ppm. Thus there appears to be a need in the United States as well as in other parts of the world for improvement in the chemical quality of municipal water supplies.

While water for agriculture does not need to meet quite as strict hygienic standards as are imposed on domestic supplies, the chemical requirements are similar to those for potable supplies. On the whole, the total dissolved solids should not exceed 1200 ppm for best results for most crops, although some crops will thrive on water containing as much as 2000 ppm. Waters containing a high concentration of sodium chloride have been shown to have a deleterious effect on some soils, causing the formation of small clusters of soil particles which are relatively impervious to water, thus impeding irrigation.

II. NORMAL FRESHWATER SUPPLIES

A. The Hydrologic Cycle

In spite of the fact that the ocean, which comprises a large fraction of the earth's water component, is too salty for use by plants and animals, it is the source of the natural potable and agricultural water supply. The process by which this is accomplished is known as the hydrologic cycle. In the course of this cycle, water is distilled by the radiant energy of nature's nuclear reactor, the sun. Its heat is absorbed by the water of the ocean, causing evaporation, as humidity, of the air above the surface. This humidified air is carried by wind circulation to regions of the atmosphere where cooler temperatures prevail. The resultant cooling of the humidified air results in the precipitation of the moisture as rain or snow. The part of this precipitation which occurs over land areas becomes the freshwater supply for terrestrial plants and animals, and replenishes the water in rivers, lakes, streams, and wells. After its use, the water and its waste return to the sea, or is evaporated directly into the atmosphere, thereby completing the cycle. Thus nature provides an effective means of desalination, and man's withdrawal of water for potable and agricultural use constitutes only a brief interruption of the hydrologic cycle.

The quality of freshwater from precipitation sources is very high, since the vapor produced is that of pure water. Its high solvent capacity, however, causes raindrops to absorb gases in the atmosphere, such as CO_2, even before they reach the ground. There is further accumulation of dissolved chemicals as the water comes into contact with soluble materials in rivers, lake beds, or in the soil through which it percolates. Water which is collected in mountain lakes or reservoirs and then transported through pipe systems is likely to have a dissolved solids content of less than 100 ppm. Water from the same source, when permitted to flow in rivers, is apt to show an increase in salinity as it progresses downstream, due both to the absorption of chemicals from the river bed and to evaporation. Water quality in rivers may be further degraded by the withdrawal of water for use and its subsequent return to the river as treated waste water. This added degradation results from the fact that the average urban use increases the salinity by 200 to 300 ppm above the water supply level.

B. Availability of Freshwater on Land Areas

The availability of freshwater on the land areas of the world is far from uniform and reflects the variability of precipitation. Great differences in precipitation appear logical when one considers the somewhat irregular geographical distribution of mountains and other topographical features, and the direction of the major air circulation, such as the trade winds. Figure 1-1 shows the relative distribution of arid and semiarid lands, classified on the basis of the amount of annual precipitation. Semiarid indicates precipitation not exceeding 25 in. per year, but more than 10 in.; arid, not exceeding 10 in. The arid area has too little rainfall to support any crops; the semiarid has enough to support certain types of crops, though certainly not all.

Inspection of Fig. 1-1 shows that, with the exception of Central Asia, a large proportion of the land within latitudes of 40° north and south is classified as arid or semiarid [4]. The amount of this water-short area is estimated to be nearly 60% of the habitable portions of the continents, and includes many areas potentially desirable for living as well as being the location of valuable mineral resources. Presently these lands support only a small fraction of the world's population. If freshwater could be supplied to these arid areas they could accommodate a substantial proportion of the increasing world population in future years. Since much of the arid land borders on oceans and seas, the obvious question is frequently raised as to whether

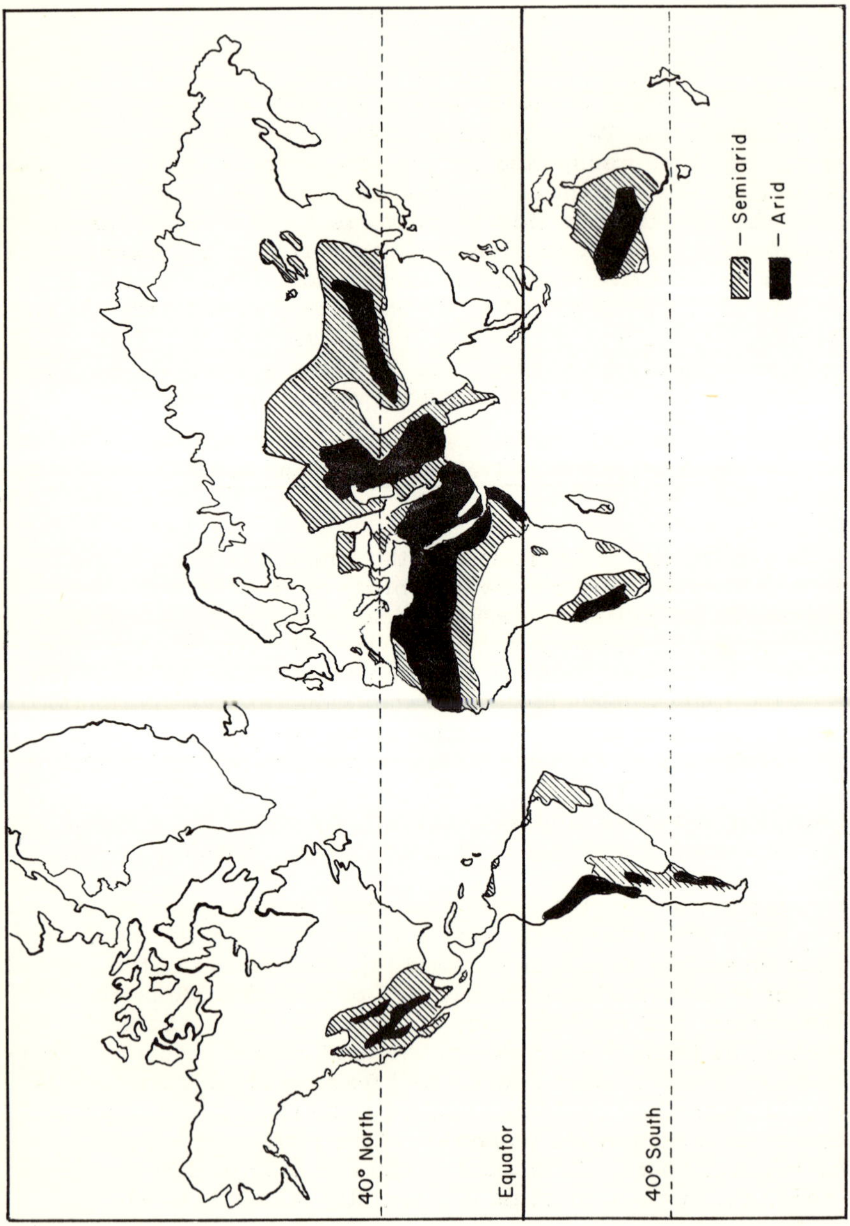

FIG. 1-1. Distribution of arid and semiarid lands [4].

freshwater can be obtained from the sea using processes other than the natural hydrologic cycle.

Not shown on the map are the numerous islands, many of which are short of water even though they may receive an annual rainfall greater than the 25 in. assigned to be the limit for semiarid regions. The shortages on such islands are due to the lack of natural water storage basins and to the short distance between the location where rain falls and where it returns to the sea. The lack of storage basins is most pronounced on the coral islands and atolls, where the maximum land elevation above the sea is only a few feet. The only natural storage on these islands is the pore space within the soil, with the result that the freshwater from rainfall sinks into the ground, forming a very thin lens-shaped pool above the level of the sea. This lens of freshwater has its thinnest edge at the beach, where the freshwater gradually discharges into the sea. The thickest part of the freshwater lens is near the center of the land mass, and it is here that "skimming wells" are generally placed to make the maximum use of the rainfall. It will be noted that the lens thickness decreases with time after rainfall, the rate of decrease being determined by the porosity of the soil, the initial lens thickness, and the width of the island. Another problem of the islands is the short length of their runoff streams, which means that water falling on slopes may be back in the sea within a matter of minutes, as compared with continents where rivers may be hundreds of miles long and the time lapse between precipitation and return to the sea may be several months. This fast runoff is as effective for causing arid conditions as too little rainfall.

III. DESALINATION AS A SUPPLEMENTARY SOURCE OF FRESHWATER

A. Freshwater Problem

Water supply problems relate to the quantity and quality of freshwater available from local sources as compared to local demand. The water required for irrigated agriculture is approximately 4 acre-feet per year for each acre of crops. The daily urban demand for freshwater in the United States averages 155 gpd per capita, of which only about 3 gpd is required for potable purposes, the difference being that required for the amenities incident to urban and industrial life. This is considerably higher than the usage in other countries. However, as standards of living rise, so does per capita consumption of water.

A very important reason that the quantity of freshwater has become a problem is the spectacular increase in population. This population "explosion" has come about through the improvement of sanitation and the achievements of the medical profession, whereby infant mortality has been decreased at one end of the life-span and death due to causes other than old age deterioration have been decreased at the other end. In spite of the effort to slow down population through birth control, the appalling population explosion makes it imperative to correct existing water shortages and to anticipate local population increases and water requirements far enough in advance that necessary construction work may be completed in phase with the demand. In many areas, local sources of freshwater have become fully utilized, making it necessary to obtain supplementary water from sources quite remote. The question then arises as to whether the transportation of freshwater over long distances will be more economical than the desalination of a local supply of saline water. This is the economic problem involving supply and demand of freshwater for an "exploding" population.

B. Development of Distillation Technology

1. Marine Usage

Distillation, the most widely used and best known process for desalting water has long been in operation in forms of varying efficiency. Distillation of water for potable purposes has been practiced on board seagoing ships for nearly 400 years. It is supposed that as early as 1593, Sir Richard Hawkins used an evaporator to produce drinking water from seawater during his long journey to the Americas. The following description published in 1758 [5] exemplifies the crude equipment used on the sailing ships of the early days.

> . . .a simple still to stand on the galley fire along with the ship's boiler. A common pot was covered with a wooden lid; in the center of the lid was bored a hole to receive a perpendicular short wooden tube which in turn was connected to another wooden tube at an acute angle. This second tube was joined to a third made of pewter by rolling up a dish and passing it obliquely through a cask of cold water to serve as a condenser . . .

This device, crude as it was, contained the central features of modern evaporators: an evaporating component and a separate condensing component.

The development of steam propulsion for ships in the nineteenth century made obsolete the stove-top unit such as just described, since steam from the main boiler was available and could be readily condensed. An example of this improved method was published in 1884 [6] and refers to the drinking water furnished to British soldiers during a military campaign in Egypt. It reads, in part, as follows:

> . . . the distilled water produced for Egypt is made in special apparatus, and various forms of condensers are employed, made under various patents. The principle involved is, however, in all cases the same. Steam is generated in one part of the ship's boilers, and condensed, filtered and aerated in a special device . . .

This article implies that seawater was fed into the ship's boilers, a practice which was generally accepted until about the beginning of the twentieth century. Prior to this time steam pressures had been low and boilers were small. The increase in steam pressures and boiler sizes, as faster and larger steamships evolved, made the use of seawater for boiler feed hazardous to both safety and continuity of operation.

The problem of boiler feed was solved by the production of distilled water, using evaporators heated by steam from the exhaust of auxiliary engines, or from the boiler itself. An early reference to the use of distilled water for steam boiler feed appeared in 1861 [7], and refers to the pioneering work on high steam pressures by Loftus Perkins and his forebears. Steam boilers and engines had been developed by them for pressures of 500 psi, necessitating the use of distilled water for boiler feed. The article reads in part:

> . . . In order to supply the place of any distilled water that might escape by leakage or otherwise, a small still should be attached to the boiler, heated by means of a coil of steampipe, of which one end communicates with the steam room of the boiler, whilst the other is over the hotwell, and is provided with a screw cock. As soon as the cock is allowed to drip or run, the still will begin to work and replenish the boiler with distilled water through the usual channel of the condenser . . .

It was some 60 years after the publication of this article before the steam pressures contemplated by Perkins found general acceptance in ship propulsion. However, the evaporation process as described by him contained the essential features of schemes for boiler feed makeup which eventually evolved for use on ships.

Because of the increased speed of ships, brought about by the development of high-pressure engines and turbines, the time at sea was decreased, so that for some voyages it seemed to be more practical to store freshwater in tanks than to produce distilled water at sea. This deterred, for a while, the impetus for improved distilled water production techniques on ocean ships. Even now most ocean-going ships carry large quantities of freshwater in tanks as well as having evaporators for water supply. The relative economy of using distilled water produced en route rather than the use of freshwater from storage tanks is under constant investigation by ship operators. The use of distilled water can be justified if the costs of owning and operating the evaporating plant are less than the income obtained by using the freshwater tank space for revenue-producing cargo.

2. Land Usage

a. Distilled Water for Boiler Feed Makeup. The use of distilled water for both potable purposes and steam boiler makeup began much later on land than on sea. This was probably due to man's inclination to choose dwelling sites and agricultural lands where natural freshwater was available. Freshwater supplies were sufficiently pure that for even boiler feed a combination of relatively inexpensive chemical treatment and frequent blowdown was satisfactory for steam boiler supply. As steam pressures increased, however, the need for purer water for boiler feed became more important. Much of the technology for producing distilled water has developed as the result of the need to provide pure water for use in steam power plants. All modern power plants now include evaporators to produce purified makeup water required for steam generation.

An example of an application of a distiller in a modern power plant is shown in the flow diagram of Fig. 1-2. In this figure it will be noted that the distiller appears near the middle section of the diagram and that it is divided into the evaporator and the evaporator condenser. The evaporator is placed in the makeup feedwater line and receives softened and preheated water from a hot-process reactor. In the evaporator this water is partially evaporated, using steam bled from the turbine as a heating medium. The vapor is passed to the evaporator condenser which is cooled by the main stream of feedwater on its way from the condenser system to the boiler, and the unevaporated portion of the water is discarded as blowdown. Condensation of the vapor for makeup water thus serves to add heat to the balance of the feedwater. Also it will be noted that the condensate from the evaporator heating steam discharges into the evaporator condenser, so that its residual heat is likewise utilized in the

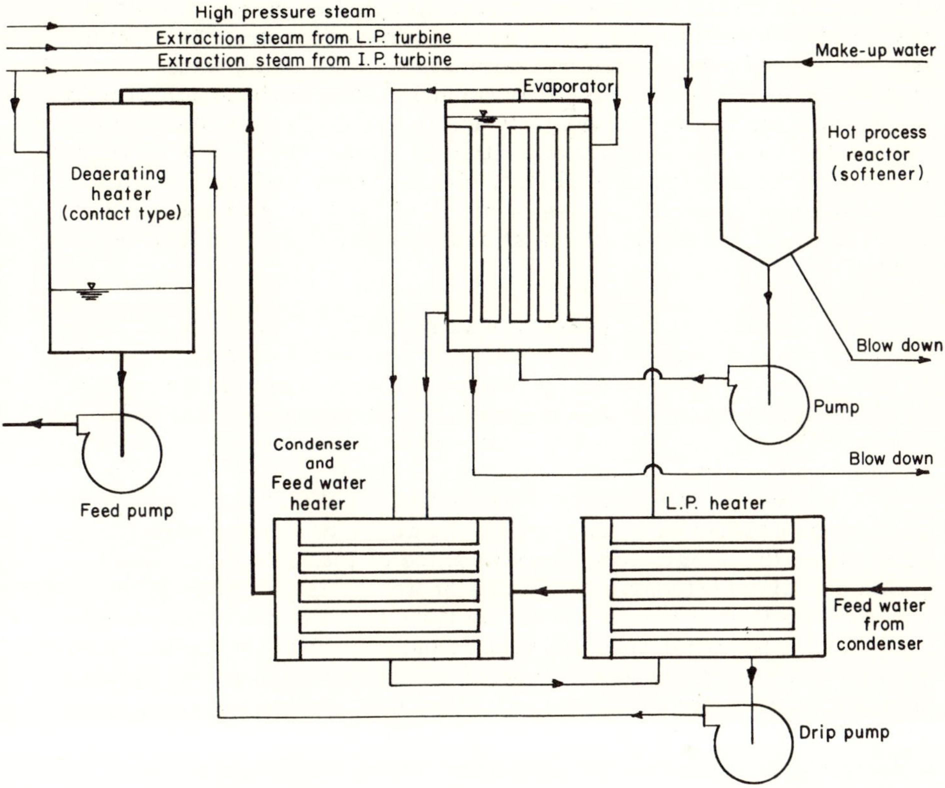

FIG. 1-2. Distiller section of modern power plant feedwater system.

heating of the feedwater. From an inspection of the figure it would be clear that the only heat escaping from the system is that associated with the liquid blowdown from the evaporator. This is a small amount of energy compared with that used in evaporating the makeup feedwater and is inconsequential in comparison. Thus the cost of heat energy required for distilling the makeup water is very small. The major concern of the designer, therefore, is not that of energy cost, but rather is directed toward high purity of product and dependability of operation.

b. Distilled Water for Potable Supplies. Techniques for producing distilled water for potable purposes have been developed in response to needs occasioned by the exploitation of mineral resources in arid lands, the poor mineral quality in local rivers or wells, and the prevention of pollution of streams. As early as 1872, it was reported that in Chile drinking water was being distilled by means of a solar device for the mules used in the nitrate mining industry. By 1930 a large number of the Chilean mining ventures were using water distilled from the brackish rivers. In other countries, the exploitation of petroleum resources in the Arabian Gulf and in the Caribbean, and the water requirements for the military activities of World War II gave increasing impetus to the development of more efficient distillation apparatus. Following World War II the use of distilled water in arid areas all over the world expanded, the cost and economy being the limiting factor.

Even as early as the 1880s it was suggested that the great incidence of typhoid in some of the cities of that time could be eliminated by the use of distilled water for drinking. Since there are no records of such installations for this purpose it is presumed that chlorination of natural waters overcame the water-borne disease problem. As late as 1926 there seems to have been little interest in distillation for potable purposes. At that date there is record [8] that the condensate from steam engines at a coal mine in Kentucky was being used for drinking by the mining community in place of the local river water which contained offensive quantities of iron salts. In many other areas of the United States reports indicate that the chemical composition of public water supplies was so undesirable that it was necessary to resort to the extensive use of bottled water, much of which was distilled. More recently distillation has become a serious consideration as a means for augmenting public water supplies.

C. Investigation of New Developments in Desalination

An extended period of low annual rainfall in southwest United States, together with the high salinities in the well waters of South Dakota and neighboring states, led the United States Congress in 1952 to initiate a program of research and development in the general field of water desalination. This gave considerable impetus to the improvement of distillation in its several forms as well as to the investigation of numerous other possible processes for obtaining freshwater from saline sources. Investigations under this program have been carried on by research laboratories, by groups in educational institutions, and by private companies under contract with the United States Department of the Interior. Independently of the federal grants, the legislature of the state of California sponsored similar investigations at the several branches of the University of California. Considerable effort has been directed toward collecting and disseminating the literature on water desalination which has been accumulating at an accelerating pace since about 1950. Special facilities for furnishing bibliographies and abstracts of this literature have been established at the Oak Ridge National Laboratory. The international magazine, "Desalination," as well as several news letters regularly chronicle research developments and other accomplishments of the desalination industry.

Similar interest in desalination is evidenced by activities in other parts of the world. In Europe, the Organization for European Economic Cooperation set up a Working Party for their study of desalination. UNESCO of the United Nations held stated meetings of its Arid Zone Program Group over a period of years, at which were included many discussions of water desalination. As a result there has come into being worldwide efforts to secure freshwater from many sources other than the usual natural local supplies.

D. Desalination as an Economic Alternative Water Source

The use of desalination is an alternative to the long distance importation of freshwater, and decisions determining its appropriateness should be made on the basis of economics. The subject of water economics is most complex, since natural water supplies for cities and agriculture are frequently commingled in origin with water used for hydroelectric power production, for flood control, and for river transportation. Control structures, such as dams and canals, are essential for all of these applications. Capital outlays for these structures are allocated to the several uses for the water and to flood control. Thus the cost of water brought from a distant watershed for augmenting a particular water supply will depend upon the relative importance of the several uses of the water from the watershed. The one cost element which is alone attributable to municipal or agricultural water supply is the cost of transporting water in a single purpose aqueduct or other conveyance.

The cost of transporting natural water supplies is highly variable, amounting in an extreme case to as much as $7.00 per kgal for well water carried 40 miles by railroad tank car and delivered to Coalinga, California. At the other end of the scale are pipelines such as the Colorado River Aqueduct serving Southern California, or the Hetch Hetchy Aqueduct serving the San Francisco, California, area, both of which deliver 200 mgd or more to the service areas at a cost of not more than $0.15 per kgal. The cost of conveyance varies inversely with the amount of water carried, i.e., the greater the quantity, the lower the cost. For small pipelines the cost may become prohibitive. In such cases, alternative methods of supply must be considered as in the case of the pipeline carrying water from mainland Florida to Key West. There, increased demand for water and deterioration of the existing pipeline over the years rendered it inadequate to deliver the additional 2.6 mgd needed at Key West. After much study of desalination as a possible alternative, it was determined that such a system could be operated at a total cost of about $0.85 per kgal, and would be less expensive than the installation and operation of a new pipeline.

The economic status of desalination as compared with long distance importation of freshwater is seldom as clear-cut as in the Key West case. There are many difficulties in assigning costs because of the confusion between cost and price, particularly when municipal ownership is involved. Very often with municipal ownership, the price at which water is sold is based only upon current operating and repair costs plus debt retirement on outstanding bond obligations. If the latter represents only a fraction of the capital investment, the resulting price may be much less than it would be if based on the amortization of the entire investment. It is important, therefore, in considerating the preparation of economic comparisons between desalination plants and alternate freshwater importation schemes, to consider comparable total costs in both cases if the comparison is to be valid.

IV. SUMMARY

Freshwater is essential for all terrestial life, and comes from the oceans via the hydrologic cycle. The uneven distribution of water on land areas has produced a large expanse of arid and semiarid lands in the middle latitudes. The population explosion has given rise to a demand for people to live in these areas, thus requiring freshwater supplies larger than local sources can produce. Desalination of local saline supplies appears to hold considerable promise for supplementing natural freshwater. Desalination has been under development for many years, first in the form of distillation, and more recently in a wide variety of processes. These processes are now under intensive development by many agencies and should be available for use in many of the cases noted above. The overriding consideration in a choice between desalination plants and long-distance importation of collected rainfall is that of economics.

REFERENCES

1. D. K. Todd, The Water Encyclopedia, 1970, Water Information Center, Port Washington, New York, 559 pp.
2. Dixey, A Practical Handbook of Water Supply, Thomas Murley & Co., London, 1931, 562 pp.
3. W. L. Patterson and H. J. Lobb, Res. Develop. Progr. Rept., No. 519, Office of Saline Water, U. S. Dept. of Interior, Washington D. C., Apr. 1970, 313 pp.
4. P. Meigs, Reviews of Research on Arid Zone Hydrology Programme, Paris, UNESCO, 1953, pp. 203-209.
5. W. Chapman, Phil. Trans. Roy. Soc. (London), 50 (2), pp. 635-9 (1758).
6. The Engineer, London, 1904, 97 (Suppl. 4), pp. 230-1.
7. A. W. Williamson and L. Perkins, The Engineer, London, 1861, and quoted by R. D. Spear, Wheeler News, 1926, 12 (No. 12), p. 4.
8. Coal Age, New York, 1929, 34 (No. 5), p. 292.

Chapter 2

PROPERTIES OF WATER AND AQUEOUS SOLUTIONS

The effectiveness of the several processes for desalting water depends upon a knowledgeable application of the properties of water and the factors which cause these properties to change. It is, therefore, essential to review the state of knowledge concerning the nature of water and its behavior under varying conditions.

I. CHARACTERISTICS OF WATER

Water has two important qualitative characteristics which distinguish it from other materials. The first of these is its stability as a chemical entity. Water is probably the most stable material known, in as much as the quantity of it available today is very nearly the same as it was many millions of years ago. It may be heated, cooled, frozen, evaporated, used to dissolve other materials, be subjected to all manner of physical and chemical change, and yet it finally returns to its normal form as liquid water. Most other materials used by man, such as metals, chemicals, fuels, and organic substances, lose their initial identity through oxidation or other change, and it is only with considerable difficulty that they return to their original state. It is this stable characteristic of water which makes possible the utilization of a wide variety of processes for separating freshwater from its many saline sources.

The second important characteristic of water is its great solvent capacity. Water dissolves many materials which may enter the solution in their solid, liquid, or gaseous states. The resulting solutions may contain completely ionized solutes, such as sodium chloride, or un-ionized solutes, such as oxygen in its molecular form. This great solvent capacity is both the source of the need for desalination and also the property which permits disposal of unwanted salts in a relatively small amount of water. One evidence of the great solvent capacity of water is the extreme difficulty experienced when very high purity water is being handled. In any chemical laboratory the distribution of distilled water to work stations by a piping system usually results in the delivery of the water to the station with a conductivity appreciably greater than that of the distilled water at the source. This increase of conductivity is due to the pickup of ions from the walls of the piping system during flow because of the high solvent capacity of the water. An understanding of these characteristics of water is of basic importance in order to deal effectively with problems encountered in the design of desalination plants.

II. THE NATURE OF WATER

When water, as a chemical entity, is compared to other compounds of hydrogen or oxygen, certain striking differences are readily observed. In the case of hydrogen comparisons, Leopold and Davis [1] note that the compounds of H_2Te, H_2Se, and H_2S have chemical formulas similar to H_2O, and that it might be expected that boiling points and other properties would vary systematically with the rising molecular weights. They found that the three former materials show regular and predictable variations of boiling point and freezing point, decreasing with a corresponding decrease of molecular weight. By extrapolation they estimated that a material with the molecular weight of water should freeze at -148°F and evaporate at -132°F if it were to fit into the pattern of the other materials. Actually, however, water freezes at 32°F and evaporates at 212°F, temperatures which are much higher and which have a range much greater than the above calculated extrapolation would suggest.

In another study Martin [2] compared water with CH_4, which has a molecular weight very close to that of water, and came to similar conclusions, namely, that the boiling point and freezing point of water are much too high, and that the range between these two temperatures is much too great to fit into any comparative analysis with other compounds. He also observed that in comparison the specific

heat of water, as well as its latent heat, is excessively large. These abnormal values for water properties suggest a need to examine the nature of water, i.e., its isotopic composition and the structural organization of the molecules in the several phases.

A. Isotopic Composition of Water

Water, a compound of hydrogen and oxygen, is known to be a mixture of several different molecular species, because hydrogen and oxygen each have several isotopes. At present three isotopes of hydrogen have been identified, together with six of oxygen. While the possible combination of these isotopes could lead to numerous different molecules, experimental observations indicate that only a small number of isotopic compounds of water actually exist under practical conditions. It is noted that molecules other than the usual $H_2{}^{16}O$, in which hydrogen has an atomic weight of 1 and oxygen of 16, appear in very small quantities. For example, the concentration of deuterium, the isotope of hydrogen with an atomic weight of 2, which serves as the hydrogen portion of the "heavy" water molecule, D_2O, is about 200 ppm and seems to be the same regardless of the source of the water. Tritium, the isotope of hydrogen with an atomic weight of 3, appears in trace quantities and is thought to originate from the effects of cosmic rays on the water molecules in clouds. This conclusion is drawn because of the fact that newly collected rainfall is found to have a maximum concentration of tritium. With its half-life of 12.5 years, the tritium content has become useful in gauging the length of time that a given sample of water has been out of contact with the atmosphere.

The known isotopes of oxygen are ^{14}O, ^{15}O, ^{16}O, ^{17}O, ^{18}O, and ^{19}O. Of these, ^{14}O, ^{15}O, and ^{19}O are radioactive with short lives and, hence, are not found in natural waters in significant amounts. For most practical purposes "pure" water may be considered to contain the following molecules: $H_2{}^{16}O$, $H_2{}^{18}O$, $H_2{}^{17}O$, HDO, and D_2O. The relative proportion of all of these isotopes, except the first, are sma[ll]. Besides $H_2{}^{16}O$ the largest concentration is $H_2{}^{18}O$, being approximatel[y] 0.2% or about 2000 ppm. In spite of the ever growing knowledge of the nature of these isotopic species, and because of their relatively insignificant quantities, "pure water" in the text which follows will be treated as containing a single component, $H_2{}^{16}O$, since this represents the overwhelming constituent of any water sample. Most al[l] analyses of water properties assume this composition for practical purposes.

B. Molecular Structure

The abnormal properties of stability and solvency of liquid water probably are related to its molecular structure and to the arrangement of two or more molecules in some repeating pattern. The concept of the structure of a single water molecule seems to have found general acceptance, but there is still wide divergence in the conception of a completely satisfactory model for the needed multimolecular pattern. However, there are certain basic facts which help to explain some of the abnormal properties of water. These will be reviewed against the background of the summary by Eisenberg and Kauzman [3] and an article by Buswell and Rodebush [4].

1. Dipolar Nature of Water Molecules

A water molecule is unsymmetrical because of the arrangement of the hydrogen and oxygen atoms. The oxygen atom, with its nucleus containing eight protons, has eight electrons in two orbital shells. The inner shell contains two electrons and the outer shell six. The electrons in the outer shell are not evenly spaced, but are so distributed as to leave two spaces into which electrons from other atoms can fit. The hydrogen atom, with its nucleus containing one proton, has only one electron in orbit around its nucleus. The water molecule is formed by the electrons associated with two hydrogen atoms entering the vacant spaces in the outer electron orbital shell surrounding the oxygen nucleus. It has been deduced that the angle made by lines joining the nucleus of the oxygen atom to the nuclei of the two hydrogen atoms as they form H_2O is 104.5°. This is very nearly the same for the isotopic molecules, D_2O and HDO. The lack of symmetry of this configuration produces a polarity of the water molecule, since the side of the molecule opposite to the hydrogen nuclei will have the negative shell exposed, while the hydrogen nuclei, which lie outside of the electron shell of the oxygen atom will exert an electrical attraction of positive sign. Thus, the water molecule becomes a sort of dipole magnet with north and south poles (see Fig. 2-1).

2. Hydrogen Bonding of Water Molecules

A second characteristic of the water molecule is related to the above, i.e., the tendency for molecules to form in clusters. This results from the polarity noted above and is usually described in terms of hydrogen bonds. These bonds are electrical forces in addition to the covalent bonds which hold the single molecule together. They

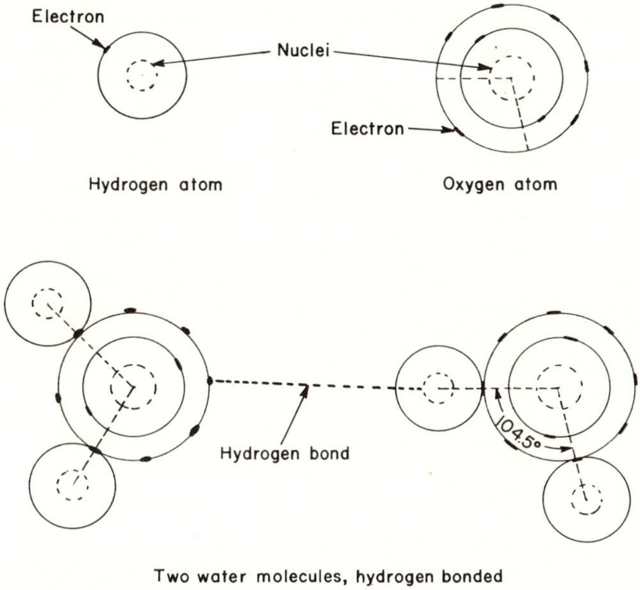

FIG. 2-1. Schematic diagram of two hydrogen-bonded water molecules and their component hydrogen and oxygen atoms. Note that the electrons are actually distributed on a spherical shell around the nucleus, although the diagrams are drawn in two-dimensional form for simplicity.

result from the fact that the single electron of the hydrogen atom can neutralize the positive attraction of its nucleus on only one side, so that the opposite side of the positive nucleus has the positive electrical force of attraction noted in the above paragraph. Thus, the positive side of each hydrogen atom in a water molecule may become bonded to an electron in an adjacent molecule. It is generally considered that each water molecule is hydrogen bonded to at least four other molecules. The result of hydrogen bonding is thought to be a very extensive and intricate network, some authors going so far as to imagine that all of the molecules in a bucket of water may be bonded to each other.

While the above description may seem to ignore the state of continual motion which obtains in all phases of matter, the conclusion would be altered very little by adding the complications involved in describing such motion. The molecules in liquid water can be pictured as being in random motion, with hydrogen bonds being either made or broken rapidly, or distorted between rather wide limits. In either case, the individual molecule can be thought to possess both translational and rotational forms of energy as well as vibrational motion between the oxygen and hydrogen nuclei within the molecule. Evidence from absorption spectra indicates that this vibration probably occurs along the lines between the oxygen nucleus and the hydrogen nuclei. There are three basic modes of vibration within the molecule: namely, one where hydrogen atoms move outward and back in phase, another where one hydrogen nucleus moves outward when the other moves inward and vice versa, and a third basic mode where the hydrogen nuclei move in directions almost perpendicular to the bonds with the oxygen nucleus. In addition, spectral measurements also give evidence that vibration occurs in the hydrogen bond between molecules. The actual vibrational motion of molecules is likely to be a blend of these four modes of vibration.

3. "Abnormal" Properties of Water

As already indicated, when the properties of water are compared with the properties of other H_2 compounds, relative to their molecular weights, it has been shown that water freezes and boils at relatively too high temperatures, has a greater temperature range between the melting point and the boiling point, and has a greater latent heat than would have been anticipated from correlation calculations. In addition to these observed "abnormalities," there are many other properties which exhibit "abnormal behavior." Although the actual observed density of water is nominally 1.0 g/cm^3, its calculated value based upon the densities of other molecules would have been expected to be 1.84 g/cm^3, assuming the molecular spacing to be the same as that measured in ice, and the molecules to be "closely packed" as in other liquids.

In a still further instance, water is unusual in that its maximum density occurs above the melting point, with the result that its crystalline form is less dense that the liquid at temperatures lower than 4°C. When subjected to increasing pressures at low constant temperatures, the viscosity of water at first decreases to a minimum value and then increases with pressure, whereas other liquids show continual increases in viscosity with increasing pressures. Furthermore, the coefficient of thermal expansion and the compressibility of water show distinct differences from other liquids when variations with temperature and pressure are considered. When these and other properties are studied, it appears that there are two important transition temperatures at which minima occur or disappear. These temperatures are 4°C, where density is maximum, and 35°C, where C_p, the specific heat at constant pressure, is minimum and above which viscosity shows no increase with increasing pressure. Transition temperatures such as these make it exceedingly difficult to set up a mathematical or structual model for the intermolecular structure of liquid water.

4. Arrangements of Molecules in Different Phases

Water in the different desalination processes may appear in gas, liquid, or solid state, the liquid state being the most important because it is the predominant form under normal atmospheric conditions, and thus is the form most available. As described above, the molecules are associated by hydrogen bonds in both the liquid and ice form. In the case of the liquid form the resulting network is one which is continually shifting as hydrogen bonds are broken with one adjacent molecule and reformed with another, or distorted without severing bonds. The thermal agitation of the molecules assures this continual movement within the network. It is conceived that the motion is quite random and it is not certain that there is any pattern of groups of molecules which is regularly repeated or duplicated in the liquid phase. In the gas phase it is generally considered that the hydrogen bonds have little or no effect, so that water vapor can be thought of as behaving similarly to the vapors of other materials.

In the case of ice, however, when water freezes it is thought that the motion of the molecules is limited to the vibrational modes, and that the hydrogen bonds hold the molecules in a distinctive, orderly arrangement. X-ray studies show this to be one in which each water molecule is hydrogen bonded to four other molecules, the five molecules together forming a tetrahedron with one molecule at the center and the other four at the vertices. The assemblage of molecules arranged thusly results in a structure with a warped hexagonal pattern, having vacant spaces or "shafts" within the hexagon. Experimental research suggests that the distance between adjacent oxygen atoms is 2.76 Å (Angstrom units), but the exact distances and locations of the hydrogen nuclei are still uncertain. This description relates to the usual form of ice, generally designated as Ice I, being the only form pertinent to desalination. There are seven other crystalline states known, but all of these can exist only at pressures greater than about 30,000 psi.

Besides ice, certain other crystalline forms of water have been used in desalination studies. These are the crystalline hydrates of certain nonelectrolyte gases. They form and melt at temperatures appreciably above 0°C, the normal freezing point of water. When examined by X-rays the structure of these hydrates are found to be cubic rather than hexagonal in basic design. The lattice structure formed by this cubic arrangement has been observed to have two different molecular spacings: the finer structure having a distance of 12 Å between molecules and the coarser one, 17 Å. Within these lattice structures are odd-shaped cavities of various sizes, those within the finer structure having either 12 or 14 wall planes and those within the coarser structure, 12 or 16 wall planes. The molecules of certain gases, such as methane, propane, butane, methyl bromide, methyl chloride, chlorine, and several synthetic refrigerants, are of sizes which exactly fill the cavities. With the cavities filled the structure is stable. The hydrate crystals are found to consist of water molecules and "guest" molecules in at least two definite ratios, depending upon the hydrating agent. The ratios observed so far are 7 and 17 moles of water per mole of hydrating agent.

The reason the hydrate crystals form and melt at temperatures higher than the normal melting point of water may be explained in terms of hydrogen bonding. The hydrogen bonds in ice are observed to hold each molecule to four near neighboring molecules, thus producing a high and uniform internal pressure. The "guest" molecules occupying the lattice cavities in the hydrate crystals do not accept hydrogen bonds, so that the water molecules on the inner surfaces of the cavities are unable to establish hydrogen bonding on the side toward the "guest" molecules. This results in these molecules having fewer than four hydrogen bonds, so that the internal pressure is less. With less internal pressure, the hydrate crystal can expand more than the ice crystal can before it reaches the point where it breaks the hydrogen bonds, or "melts." Therefore, this added expansion capability means that the temperature can be higher for the melting of hydrate crystals than for ice. Experimental data of hydrate melting points show propane hydrate crystals to melt at 5°C and crystals of methyl chloride hydrate to melt at 21°C, as compared with ice at 0°C.

5. Some Effects on Properties

Hydrogen bonding and the dipole nature of the water molecule can be used to explain some of the characteristics of water, one example being the solvent capacity. When a substance, such as sodium chloride, is dissolved in water, each ionic component of the salt is immediately surrounded by water molecules held in a cagelike formation by the hydrogen bonds. The dipolar nature of the water molecule results in the molecules surrounding the positive sodium ions being oriented with the oxygen nuclei inward, while those around the negative chloride ions have the hydrogen nuclei inward. The residual electrical charges on the water molecules partially neutralize the valence bonds between the sodium and the chloride ions, and the molecular "cages" prevent the two ions from coming very close together and reuniting the ions into a molecule of sodium chloride. This capacity to keep the ions apart is the key to the effectiveness of water as a solvent.

Hydrogen bonding also accounts qualitatively for phenomena such as the high surface tension of water and the large latent heat of vaporization. In the case of surface tension, exhibited by the liquid level rise in a capillary tube, hydrogen bonds on the molecules next to the wall reach out to oxygen atoms in the tube wall surface, i.e., the oxygen in the SiO_2 if the tube is made of glass. The formation of hydrogen bonds between water molecules and oxygen atoms in the wall pulls the water surface upward until the gross weight of the water column equals the sum total of the hydrogen bonds between the surface molecules of water and the oxygen atoms in the wall.

In the case of vaporization, the hydrogen bonds between molecules must be broken if a molecule is to escape from the surface. Since each water molecule is attached to its neighbors by at least four hydrogen bonds, its escape from the surface as vapor involves considerable energy, this energy being represented by the latent heat. When it is noted that the molal latent heat for water is nearly twice that of materials such as ammonia and hydrogen sulfide, it seems clear that the effect of hydrogen bonding is appreciable.

III. PROPERTIES OF PURE WATER

The properties of pure water which have been found relevant to calculations pertaining to desalination processes include the following:

1. Engineering thermodynamic properties, such as pressure-volume-temperature relations, heat capacity, enthalpy, and entropy

2. Physical and thermal properties involved in fluid flow and heat flow, such as viscosity, density, and thermal conductivity

3. Miscellaneous properties, such as activity, electrical conductivity, and surface tension

The rather incomplete knowledge of the structure of water, as cited in the previous section, makes it necessary to depend on semiempirical data for numerical values. The designation, semiempirical, is used in recognition of the fact that many of the properties of water are interrelated, and the consistency of independently determined experimental values can be established by thermodynamic reasoning and calculations. Experimental determinations of the values for properties generally assume the use of distilled water, free from gases, and exhibiting a very low electrical conductivity. Although not measured, the normal amounts of "heavy" water and other molecular species were probably present, so that the properties measured are those of an isotopic mixture. However, the amounts of isotopes other than normal water are so small that it is likely that the values obtained for properties are very close to those which would be obtained for $H_2{}^{16}O$ alone.

The experimental data basic to the engineering thermodynamic properties include pressures, volumes, temperatures, heat capacities, and rates of change of volumes and of heat capacities with changes in pressure and temperature. Experimental determinations utilize a number of procedures and techniques, including calorimetric measurements, and the determination of the pressure-temperature relations under saturated conditions. These data are correlated by the use of thermodynamic interrelations, and interpolation equations are formulated for enthalpy, entropy, and specific volume. Through the comparison of results by different experimenters at a series of International Steam Table Conferences, the thermodynamic properties of steam and saturated water have been determined with a high degree of precision and are published in tabular form in numerous books. In the United States, the steam tables most widely used are those by Keenan and Keyes [5] and the ASME [7]. The Keenan and Keyes values are also published in tabular form in the OSW Data Book [8] and in graphic form by Ellenwood and Mackey [6].

The properties listed in items 2 and 3 of the first paragraph of this section have been measured by various experimenters, and the resulting data have been summarized by several reviewers. It is generally found that numerical values of these properties are less precise than those of the engineering thermodynamic properties noted above, partly because of experimental difficulties but primarily because of the absence of thermodynamic interrelations. Thus it may be expected that values of these latter properties will be modified from time to time as experimental techniques are improved. A satisfactory summary of these properties appears in graphical form in the OSW Data Book [8] and represents the best values available in 1964-1965.

IV. PROPERTIES OF AQUEOUS SOLUTIONS

It is important to compare the above groupings of properties of pure water with properties of aqueous solutions. Seawater and other saline water from which freshwater may be obtained fall into the general category of aqueous solutions. Many of the calculated values of the properties of these solutions may be obtained by applying correction factors based on the concentration and composition of the solution. Some of these correction factors are for boiling point rise, freezing point depression, and specific gravity, all relative to the basic values for pure water at the same temperature. Tabular and graphic values for these and other properties appear in the OSW Data Book [8] for standard seawater, for sodium chloride solutions, and for magnesium chloride solutions. These were arrived at experimentally. It would be convenient if one could calculate these values by using the simple laws of physical chemistry pertaining to "ideal solutions," and thus obtain values for boiling point rise, etc. This is impossible, however, for it has been found that calculated values so obtained for seawater, which has a concentration of approximately 0.5 moles per liter, are quite different from those obtained experimentally. For more dilute solutions, some of which have concentrations as low as 0.02 moles

TABLE 2-1

Salinity of Water in Various Seas

	Concentration (ppm)
Baltic Sea	7,000
Black Sea	13,000
Adriatic Sea	25,000
Pacific Ocean	33,600
Indian Ocean	33,800
Atlantic Ocean	36,000
Mediterranean Sea	39,400
Arabian Gulf	43,000
Red Sea	43,000

TABLE 2-2

Ionic Composition of Seawater

	Ref. [8] ppm	Ref. [9] ppm	Ref. [9] Equivalents per kg of water
Cations			
Sodium, Na^+	10,556	10,561	0.47560
Magnesium, Mg^{2+}	1,272	1,272	0.10833
Calcium, Ca^{2+}	400	400	0.02068
Potassium, K^+	380	380	0.01007
Strontium, Sr^{2+}	13		
Total	12,621	12,613	0.61468
Anions			
Chloride, Cl^-	18,980	18,980	0.55438
Sulfate, SO_4^{2-}	2,649	2,758	0.05946
Bicarbonate, HCO_3^-	140	—	—
Fluoride, F^-	1.1		
Bromide, Br^-	65	65	0.00084
Total	21,834	21,803	0.61468
Boric acid, H_3BO_3	26		
Total Dissolved Solids	34,481	34,416	

per liter, the "ideal solution" concept has been found to be applicable for the computation of certain of the properties.

A. Seawater

The properties of seawater have received major attention, partly

TABLE 2-3

Molecular Composition of Seawater

	Concentration	
	ppm	moles/kg
NaCl	23,476	0.40100
$MgCl_2$	4,981	0.05230
Na_2SO_4	3,917	0.02760
$CaCl_2$	1,102	0.00994
KCl	664	0.00890
$NaHCO_3$	192	0.00228
KBr	96	0.00081
H_3BO_3	26	0.00042
$SrCl_2$	24	0.00015
NaF	3	0.00007
Total	34,481	0.50347

because of its relative chemical uniformity and abundance and also due to its use in a majority of the existing desalination plants. Seawater has been analyzed in many parts of the world and has been found to have a fairly uniform combination of chemicals in most of the seas. The total concentration of dissolved salts, on the other hand, varies appreciably as is evidenced in the partial list in Table 2-1, in which the values of total dissolved solids in seawater are seen to vary from 7000 to 43,000 ppm. Although the published composition of "standard seawater" varies slightly from one author to another, there is a great deal of similarity between the several versions. Over forty materials have been identified in seawater, but only eleven ions appear with concentrations of 1 ppm or more. These latter ions and their concentrations in "standard seawater" [8] are given in the first column of Table 2-2. The corresponding molecular composition of "standard seawater" is shown in Table 2-3.

Since the carbonate and bicarbonate ions tend to disintegrate upon heating, thus interfering with the precise experimental determination of properties, it is common practice to use seawater solutions which

TABLE 2-4

Constants for Determining Boiling Point Rise[a]

Concentration ratio	a	b
0.4	0.113	0.001193
1.0	0.308	0.003092
1.4	0.453	0.004442
2.0	0.696	0.006579
2.4	0.873	0.008083
3.0	1.165	0.01045
3.6	1.482	0.01297

[a]Concentration ratio = concentration of sample to normal seawater. a and b are computed from data of Fabuss and Korosi [9] for use in Eq. (1). (Valid from 60° to 300°F.)

lack these materials. The seawater solution employed in many studies of properties are often synthetically made up by dissolving known weights of the several chemicals in distilled water, with the carbonate and bicarbonate ions replaced by sulfate ions on a stoichiometric basis. An example of the formulation of such a carbonate-free seawater solution is that used by Fabuss and Korosi [9] in determining densities, boiling points, etc., and is given in terms of its ionic percentages in the second column of Table 2-2, and in gram equivalents per kilogram of pure water in the third column.

Using this saline solution, Fabuss and Korosi determined the boiling point rise for various seawater concentrations. The results of their work can be represented algebraically by the linear relation:

$$\text{BPR} = a + bT = \text{boiling point rise in }^\circ\text{F} \qquad (1)$$

where T is temperature in °F; and a and b are constants for any given concentration of seawater.

The concentration of seawater is expressed as a concentration ratio, i.e., the ratio of the total dissolved solids at a given concentration to the total dissolved solids of "standard" seawater (34,416 ppm, column 2, Table 2-2). Table 2-4 gives the numerical constant values of a and b for the several concentration ratios covered by the Fabuss and Korosi data. As an example of the results that can be obtained by the use of this analysis, the value of the boiling point rise of "standard seawater", computed on the basis of Eq. (1), is 0.9639°F at 212°F. However, for most of the other properties of seawater, rather than using calculated values, it is usual practice to obtain approximate values from the OSW Data Book [8] where they are displayed in both tabular and graphic form. It will be noted that these charts show the property values for various concentrations of seawater. This is important, not only because seawater concentrations vary according to their source, but also because they vary throughout a plant as a result of the removal of pure water by the desalination process.

B. Other Saline Water Sources

As compared with seawater, saline waters from wells, lakes, and streams are highly variable, not only in the total concentration of dissolved materials but also in the combinations of chemical ions. This is logical when one considers the great variability of geological structure with which different water samples have been in contact. The diversity of ionic composition of natural saline waters is illustrated by the values given in Table 2-5 for five samples from different sources. The table shows only the principal ions whose presence and amounts can be determined by relatively simple tests. The presence of trace amounts of other ions, e.g., fluorides and heavy metals, is not specified in this table because the quantities are so small that they have a negligible effect on the properties of interest in desalination plants. It is only in cases where trace amounts of these common ions would be objectionable that the detailed analysis of Table 2-5 is an important consideration.

The water samples listed in Table 2-5 are seen to contain some of the same ions that are found in seawater, as shown in Table 2-2, but in different proportions. A comparison of the components of these five samples with each other and their total with that of seawater indicates considerable variation in the ratios of the several ions, e.g., the ratio of the ppm of sodium ions to the ppm of calcium ions. For plant design purposes it is possible to deduce certain approximations for values of the properties of these different aqueous solutions without resorting to extensive experimental work. For solutions having concentrations of not more than 0.02 moles per liter, the simple rules pertaining to ideal solutions are applicable. For more concentrated solutions, values may be estimated from properties of solutions having

TABLE 2-5

Ionic Composition of Brackish Water

Dissolved ions	Coalinga California (well)	Utah (well)	Reference water[a] No. 1	Reference water[a] No. 2	Reference water[a] No. 3
	1	2	3	4	5
Cations, ppm					
Sodium, Na^+	500	542	1170	494	74
Calcium, Ca^{2+}	120	102	148	65	204
Magnesium, Mg^{2+}	72	54	47	40	127
Potassium, K^+	9				
Iron, Fe			1	2	
Manganese, Mn^{2+}				1	
Boron, B	3				
Total	704	698	1366	602	405
Anions, ppm					
Chloride, Cl^-	252	1090	2040	72	50
Sulfate, SO_4^{2-}	1170	38	137	1060	984
Bicarbonate, HCO_3^-		110	96	260	159
Carbonate, CO_3^{2-}					
Nitrate, NO_3^-	0.8			4	
Fluoride, F^-	0.2				
Silica, SiO_2			25		
Total	1423	1238	2298	1396	1193
Grand total	2127	1936	3664	1998	1598
pH		7.8	7.1	7.4	7.2

[a]Used for testing electrodialysis units [12].

the same molal concentration, as for instance, diluted seawater, sodium chloride solutions, or calcium chloride solutions. These three aqueous solutions have been studied extensively and their properties are summarized in the OSW Data Book [8]. Numerical values so obtained constitute the practical means of approximating pertinent property values for these relatively dilute aqueous solutions from specific saline sources.

It should be realized that the magnitude of certain differential properties, such as boiling point rise, are so small for these dilute solutions that for practical purposes they may be omitted. For instance, a water having a salinity of 2127 ppm, as at Coalinga, has a concentration of about 0.05 moles per kg. An ideal solution of this concentration would have a boiling point rise of 0.04°F, based upon a value of 0.94°F per mole of concentration. Since 0.04°F is so small as to have negligible effect on heat flow calculations, it can be neglected.

In contrast, properties such as osmotic pressure and electrical conductivity are quite sensitive to small changes in the state point of water and in its burden of dissolved materials. Further information will be presented in later chapters, where specific desalination processes are being discussed.

C. Scale-Forming Tendencies of Aqueous Solutions

Waters which contain ions capable of forming molecules of materials which are only slightly soluble in water may become supersaturated at some point in a desalination process. Supersaturated materials tend to precipitate if nuclei are present and the supersaturated condition persists long enough for the precipitation reaction to be completed. Precipitation next to a solid surface may be promoted by minute irregularities on the surface, which serve as nuclei, and by mass or heat transfer at the surface which increases the local concentration of the dissolved materials. Precipitation next to a solid surface generally produces a hard and adherent layer of the chemical which drops out of the solution. This layer is referred to as "scale," and is encountered in most processes used for desalination. Scales can form on heat-transfer surfaces in all types of distillation plants thereby interfering with heat flow between the surface and the liquid as well as with the movement of the liquid past the surface. It also offers problems in other types of plants, such as in electrodialysis, where the scales interfere with the circulation of water through the very narrow passages used in such equipment.

While there are special circumstances under which other materials may form scales, the ions common to most waters and which appear in the chemical analysis of most scales are calcium, magnesium, sulfate, and carbonate. An inspection of the water analyses in Tables 2-2 and 2-4 show that these ions are quite common. Bicarbonate ions

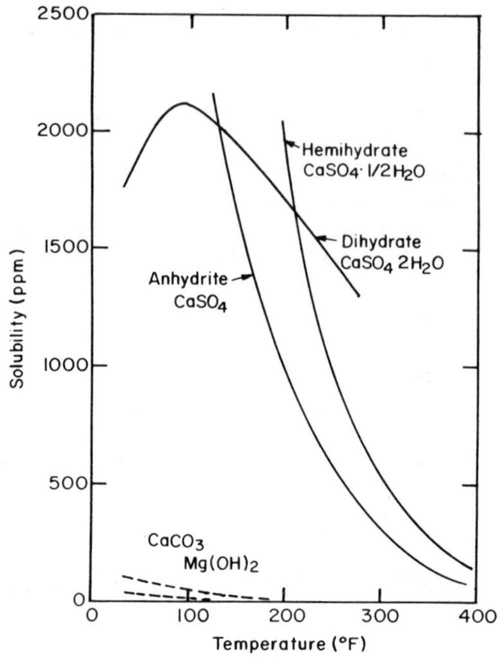

FIG. 2-2. Solubility of scale-forming chemicals in pure water.

are also present in some land-originated waters as well as in seawater. The most common compounds of materials found in scales are calcium carbonate, magnesium hydroxide, and calcium sulfate, the latter of which has been identified in at least three crystalline forms: the anhydride $CaSO_4$, the hemihydrate, $CaSO_4 \cdot \frac{1}{2} H_2O$, and the dihydrate, $CaSO_4 \cdot 2 H_2O$.

The compounds listed in the above paragraph are only slightly soluble in pure water and, unlike nonscaling compounds, have solubilities which decrease as temperature is increased. The variations of solubility with temperature are shown in Fig. 2-2 in which the concentration limit of solubility in pure water of each of the several salts is shown as a function of temperature. In each case the curve shows the concentrations above which the single-component solution would be supersaturated and would tend to produce scale. Also from this curve it is noted that only some 50 ppm of $CaCO_3$ will dissolve in pure water, while the components dissolved in seawater amount to over 100 ppm. From this it is clear that the solubility of a chemical in complex aqueous solutions, such as seawater, may be quite different from that in distilled water. Lacking mathematical equations and because of this complicated distinction between the solubility of salts in pure water and in seawater, the prediction of ionic concentrations at which scales will tend to form has had to be based primarily on empirical data. Inasmuch as $CaCO_3$ does not precipitate from normal seawater at ambient temperatures, but only at high temperatures, it is important to consider the reasons for this and to establish, if possible, a scheme for predicting its solubility limits as well as other scale-forming compounds in solutions of various compositions and concentrations. Also, in addition to temperature influence, the precipitation of $CaCO_3$ and $Mg(OH)_2$ is strongly affected by the pH factor of the solutions, i.e., by the degree of alkalinity. In contrast to this, the precipitation of $CaSO_4$ and its hydrates is relatively independent of pH. For this reason the solubilities of $CaCO_3$ and $Mg(OH)_2$ are considered separately from that of the calcium and sulfates.

1. Alkaline Scale Compounds

The predominant ionic form basic to the formation of the two alkaline scale compounds, $CaCO_3$ and $Mg(OH)_2$, is the bicarbonate ion, HCO_3^-, common in most waters. When heated to temperatures of about 160°F this ion tends to disintegrate according to the relationship:

$$2\ HCO_3^- = CO_3^{2-} + CO_2 + H_2O \tag{2}$$

With the removal of the CO_2 gas from the solution, the CO_3^{2-} ions are left to combine with calcium ions to form $CaCO_3$ when conditions are favorable for its precipitation. Any carbonate still remaining will disintegrate on further heating according to the equation:

$$CO_3^{2-} + H_2O = CO_2 + 2\,(OH)^- \tag{3}$$

The resulting CO_2 gas escapes from the solution, permitting the OH^- ions to combine with magnesium ions to form the very slightly soluble $Mg(OH)_2$. Thus both of these scale components result from the disintegration of the bicarbonate ion.

Both these precipitates, $CaCO_3$ and $Mg(OH)_2$ are soluble in acids; thus scales containing these two compounds are readily dissolved by adding acid to the feedwater. For example, in a seawater solution a change from pH 8 to 7 will increase the solubility of $CaCO_3$ by a factor of 10, dissolving much of the precipitate which otherwise would have formed. This acid is effective not only in dissolving these scale components but also in preventing their formation, as is shown in the following equation:

$$H^+ + HCO_3^- = CO_2 + H_2O \tag{4}$$

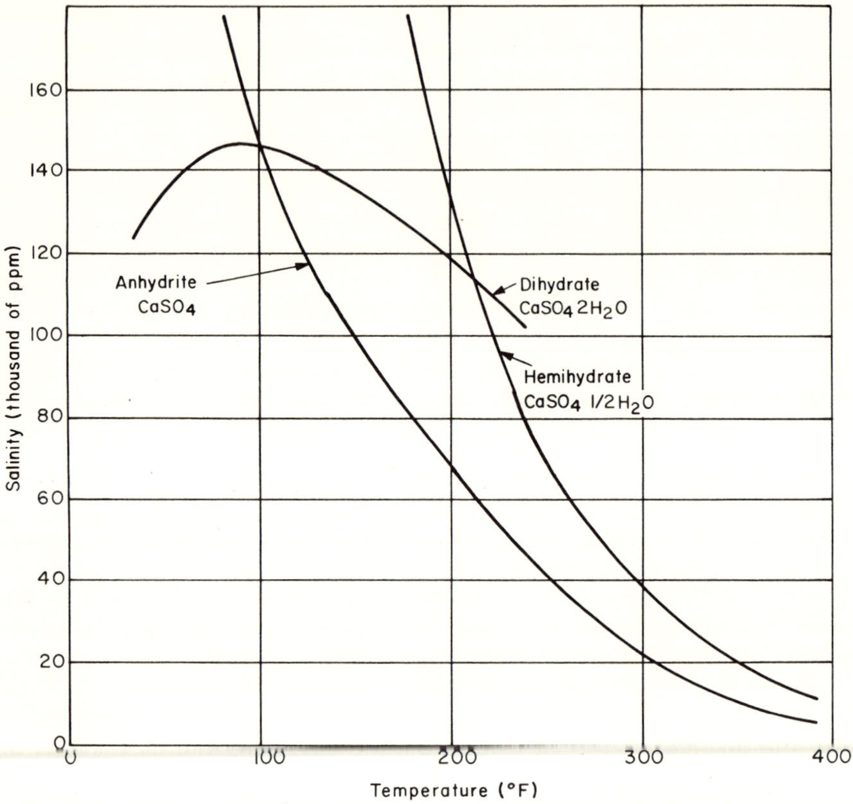

FIG. 2-3. Solubility of scale-forming chemicals in seawater.

Pretreatment, based on this concept, is used in many distillation plants and involves acidification of the feedwater in an amount just sufficient to neutralize the bicarbonate ion concentration. There must also be a scheme for degassing the CO_2 from the system. Futhermore, care must be taken not to overacidify, since corrosion problems can result from low pH. In some plants, an alkali such as NaOH is added after degassing in order to restore a favorable acid-alkali balance.

2. Sulfate Compounds

Aqueous solutions from which the bicarbonate ion has been removed by acid treatment may contain calcium and sulfate ions. Under certain conditions of temperature and concentration these ions tend to combine and deposit as a calcium sulfate scale. Whether such scale will form depends primarily upon the solubility limits of the scale-forming ions in the particular solution. The solubility limits for calcium sulfate and its hydrates have been determined experimentally for seawater in which the bicarbonate ion has been replaced with a sulfate ion. This modified seawater has been used in these experiments because the vast majority of distillation plants use dosage of the feed with sulfuric acid, followed by the removal of the CO_2 gas, described in the above equations, for the purpose of eliminating the problem of alkaline scale. Basic experimental data, presented by Langelier et al. [10] in 1950, were used for a number of years as a guide for scale treatment. Lu and Fabuss [11] reviewed this early data in the light of later publications and of plant experience. They were able to point out inconsistencies in the accumulated information, and developed a revised scheme for predicting solubility limits; a scheme which they showed to be verified by experimental results.

The scheme of calculation presented by Lu and Fabuss utilizes the principle that the solubility product is nearly constant for a slightly soluble salt in the presence of an added salt having one ion in common. In the case of calcium sulfate in seawater, the common ion is sulfate, which is present in excess of the amount needed to combine with all of the calcium, and, hence, represents a second dissolved salt. The calculation of the solubility product requires the estimation of the "activity coefficient," obtained from a modified Debye-Hückel formulation. This formulation, involving the ionic strength of the solution, contains empirical constants based on available expermental data. The results of the Lu and Fabuss investigation for seawater are shown in Fig. 2-3. The article cited also includes the constants for solutions of $MgCl_2$ and NaCl, making it possible to estimate the limits of solubility of calcium sulfate in aqueous solutions other than seawater.

An examination of Fig. 2-3 indicates that serious scale-forming problems can be expected to occur in plants distilling seawater when the brine temperature exceeds about 250°F. If it is desirable to operate such plants at these higher temperatures, some means of reducing the scale forming tendencies must be found, the most obvious being partial or complete removal of either calcium or sulfate ions. Further discussion of this point will be found in the chapter on Distillation.

V. SUMMARY

The study of desalination processes requires a knowledge of the properties of water and aqueous solutions. Water differs from other natural liquids in being extremely stable, having a very high solvent capacity and having freezing and boiling points much higher than is consistent with other compounds of hydrogen. These unusual properties may be rationalized in terms of the polar nature of the water molecules.

Water is a mixture of several compounds of hydrogen and oxygen, the result of different combinations of the several isotopes of these two elements. Compounds other than ordinary water appear in quantities sufficiently small that they have little effect upon the properties of water.

Hydrogen bonding between adjacent molecules of water follows a definite geometric pattern in the crystalline forms of ice and gas hydrates. In the liquid phase, hydrogen bonding does not produce a regular structural pattern, but exerts a strong influence on solvent capacity, surface tension, latent heat, and other properties.

The engineering thermodynamic properties of pure water, as given in published steam tables, are known with a high degree of precision, but certain other properties have values which are completely empirical, and are not subject to the thermodynamic interrelations of the first group.

The properties of aqueous solutions, including seawater and saline waters used for desalination, are expressed in part in terms of their deviations from the properties of pure water—an example being the boiling point rise of an aqueous solution. For solutions as concentrated as seawater, these properties must be determined experimentally, since the simple laws of ideal dilute solutions do not seem to apply.

In addition to the properties which affect the performance of a desalination plant, seawater and natural saline waters contain dissolved ions which tend to form scale on solid surfaces and thus interfere with the functioning of the plant. A knowledge of the solubility limits of the scale-forming compounds and methods for predicting these limits is essential if scale formation and its attendent problems are to be avoided.

REFERENCES

1. L. B. Leopold and K. S. Davis, Water, 1966, Time Inc., New York, 200 pp.
2. D. F. Martin, Marine Chemistry, 1970, Marcel Dekker, New York, 447 pp.
3. D. Eisenberg and W. Kauzman, The Structure and Properties of Water, 1969, Oxford University Press, New York, 296 pp.
4. A. M. Buswell and W. H. Rodebush, Sci. Amer., 1956, 194 (No. 4), pp. 76-89.
5. J. H. Keenan and F. G. Keyes, Thermodynamic Properties of Steam, 1936, John Wiley & Sons, Inc., New York, 89 pp.
6. F. O. Ellenwood and C. O. Mackey, Vapor Charts, 1939, John Wiley & Sons, Inc., New York, 42 pp.
7. ASME, 7th Intern. Conf. Properties Steam, 1970, American Society of Mechanical Engineers, New York, 239 pp.
8. M. W. Kellogg Co., Saline Water Conversion Engineering Data Book, 1965, U. S. Dept. of Interior, Washington, D. C.
9. B. M. Fabuss and A. Korosi, Office of Saline Water Res. and Development Prog. Rep., No. 384, 1969, U. S. Dept. of Interior, Washington D. C.
10. W. F. Langelier, D. H. Caldwell, and G. F. Lindholm, Final Report to Engineer Research and Development Laboratories, Contract DA-44-009-Eng-193, Institute of Engineering Research, University of California, Aug. 15, 1952.
11. C. H. Lu and B. M. Fabuss, Ind. Eng. Chem. Process Design Develop. 7 (No. 2), April, 1968, pp. 206-212.
12. A. P. Christodoulou, G. R. Olsson, and H. J. Monnik, Parametric Economic and Engineering Evaluation Study of the Electrodialysis Process for Water Desalting, Research and Development Prog. Rep., No. 488, 1969, Office of Saline Water, U. S. Dept. of Interior, Washington D. C.

Chapter 3

GENERAL ENGINEERING CONSIDERATIONS

In providing a desalination facility the tasks of the engineer are to design the plant so that it will produce the required amount of high quality water and to predict the performance of the plant and the cost of the water produced. The traditional restraint on the work of the engineer has been the economic competition of alternative schemes for accomplishing the same objectives. Recently new constraints have been added, so that the engineer is now expected to produce plants which are harmonious with the environment and which will produce little or no pollution. Under these circumstances the engineer concerned with the design and construction of desalination plants is confronted with a formidable task. He must produce desalted water at a cost which is competitive with the cost of water transported from a mountain catchment area, realizing that the desalting plant cannot have the aesthetic attractiveness of a reservoir in the mountains, but yet must be made to be ecologically acceptable. He further is aware that the energy required by any desalination plant carries with it some waste heat, and that the process itself produces a relatively concentrated waste brine. Disposal of the waste heat and brine presents serious problems insofar as pollution is concerned. This chapter deals with some of the problems of engineering design and of pollution control which are inherent in all methods of desalination.

I. ANALYSIS OF COSTS

A. Cost Components

The overall consideration in deciding upon the type and size of a given water production process is an analysis of its costs. These depend upon the interrelations of the several components of water cost which include the following:

1. Costs due to capital amoriization of the desalination plant equipment
2. Costs due to heat energy used for water production and/or its equivalent
3. Costs due to electrical energy required for operating pumps and other auxiliary machinery
4. Costs for supervision, maintenance, and operation, including both labor and materials
5. Costs related to site and buildings
6. General overhead costs

The last three items are strongly influenced by location and by factors not necessarily directly related to the desalination equipment, and therefore cannot be treated on a general basis. In contrast, the first three elements are related to the basic process and design of a desalination plant and, for general purposes, can be optimized independently of the last three elements. Analysis of these components and computation of their values constitute a basic requirement for plant design engineers.

1. Costs due to capital amortization of any desalination plant can be expressed in terms of the capital cost per unit area of heat or mass transfer area used, the productivity per unit area, and the rate of amortization. Expressed in other words, the cost per thousand gallons due to capital investment is the annual cost of the investment (including depreciation, interest on the money, sinking funds, etc.) divided by the annual production of the plant. Expressing this in symbols, the relationship is as follows:

$$C_1 = 8333\,apA/(W \times 24 \times 365) \tag{1}$$

where C_1 is amortization cost per kilogallon, a is the cost of plant per unit area, in dollars, p is fractional charge per year, A is the total production area of plant in square feet, W is the production rate in pounds per hour, and 8333 is the weight of water, lb/kgal. Therefore

$$C_1 = \frac{0.9513\,ap}{(W/A)} \tag{1a}$$

2. Costs due to the energy used in the separation of the freshwater from the saline source can be expressed in several ways. Where electrical or mechanical energy is used directly, as in electrodialysis or reverse osmosis, this cost element is readily determined:

$$C_2 = 8333\,Eg \tag{2}$$

where E is the energy required in kWh/lb and g is the cost of power, dollars/kWh.

Where steam is used, as in distillation, the expression in terms of the cost of steam is as follows:

$$C_2 = \frac{8.333 \times 10^{-3}\,bh_{fg}}{(W/W_s)} \tag{2a}$$

where b is the cost of steam, dollars/10^6 Btu, h_{fg} is the latent heat of steam, Btu/lb, and W/W_s is the GOR, lb distillate/lb steam.

3. Costs due to auxiliary electrical energy, the major component of which is that required for pumping the water through the plant. The cost of energy per kilogallon for the work of pumping has the general form:

$$C_3 = \frac{(DfV^3 g/e)}{(W/A)} \tag{3}$$

where f is the friction factor, g is the cost of power, V is the velocity of water through passage, e is the efficiency of the pumping machinery, W/A is the production rate, lb/ft^2/hr, and D is the constant to convert units into dollars/kgal.

The total cost of water, therefore, due to capital amortization and to the energy required, is the sum of the above three components:

$$C^* = \frac{0.9513\,ap}{(W/A)} + \frac{8.333 \times 10^{-3}\,bh_{fg}}{(W/W_s)} + \frac{(DfV^3 g/e)}{(W/A)} \tag{4}$$

Equation (4) is seen to have two terms involving the rate of distilled water production per unit area, (W/A), and one involving the GOR, (W/W_s), or the energy economy. It should be emphasized that the last term in the equation is of low magnitude except for certain specific types of plant; for instance, it is unimportant in most types of distillation except multiple-stage flash plants.

B. Optimization of Costs

The objective in the design of a plant of a given capacity for a specific location is generally to produce freshwater at minimum cost. As shown in the previous section, the two major components of water cost, inherent in all plants, are the amortization of the capital investment and the charge for energy. The net charge for energy can be reduced by increasing the efficiency of the plant. Since increased efficiency is generally accomplished by an increased capital investment, it is evident that minimum cost of the product water will probably result from a compromise design which is neither the most efficient in its use of energy nor the cheapest possible from the investment point of view.

To arrive at the optimum design it is necessary to explore the effects of changing the values of the several different parameters characteristic of a given type of desalting plant. While this was formerly a most formidable task, sometimes requiring thousands of man-hours of work using slide rules, plotted curves, and electric calculating machines, it is presently possible to optimize a design in a relatively short time by using appropriate programs on digital computers. The ranges of some of the quantities to be varied during an optimization study can be reduced if one accepts the basic assumption that minimum water cost will be realized when the energy cost is made approximately equal to the cost of amortization. The following arguments demonstrate the validity of this assumption.

Inserting numerical values in the numerators of the first two terms of Eq. (4) results in an equation of the form:

$$C^* = \frac{R}{(W/A)} + \frac{S}{(W/W_s)} \tag{5a}$$

where R and S are numerical coefficients. Since W_s/W is proportional to the energy input E, Eq. (5a) can be rewritten as:

$$C^* = \frac{R}{(W/A)} + S'E \tag{5b}$$

where S' includes S and the conversion factor for energy units.

In seeking to optimize any plant producing a given quantity of water, W, the area of heat or mass transfer surface, A, and the rate of energy input, E, are the values that must be varied by changing the plant design. It is obvious that increasing A generally decreases the energy requirement, E, the ultimate minimum E being reached when the area becomes very large and approaches infinity. This theoretical limit of energy requirement represents the ideal energy of separation, or the lowest possible value required to produce the distillate. By this reasoning the quantity (W/A) can be replaced with some function of E, the simplest one being a straight proportionality such as $W/A = \text{constant}(E - E_0)$, where E_0 is the ideal energy of separation. Rewriting Eq. (5) with this substitution gives the result:

$$C^* = R''/(E - E_0) + S'E \tag{6}$$

where R'' includes all conversion factors needed to give the correct units plus the value of W for the plant.

To determine the minimum value of water cost, it is possible to differentiate the terms on the right-hand side of Eq. (6) with respect to E and equate the result to zero, as in the following:

$$\partial C^*/\partial E = -R''/(E - E_0)^2 + S' = 0$$

or

$$E - E_0 = (R''/S')^{\frac{1}{2}} \tag{7}$$

Substituting this result into Eq. (6) gives the minimum cost of water as:

$$C^* = (R''S')^{\frac{1}{2}} + S'E_0 + (R''S')^{\frac{1}{2}} \tag{8}$$

For even the most efficient plants constructed to date the energy required is many times the ideal minimum energy, so that the middle term on the right-hand side of Eq. (8) is very small compared with the other two. Since the other two terms are equal, it follows that the minimum cost will result when the cost of amortization is very nearly equal to that of the energy involved in the process. This general conclusion is borne out by the cost relationships observed in existing plants.

II. MINIMUM ENERGY REQUIREMENTS

The important influence of the energy requirement on the cost of the product water raises the question as to whether a given process has been developed to approach its ultimate maximum economy. If not, it would be useful to know the value of the ultimate economy, and the conditions which could lead to its realization. While it must be self-evident that no process has reached the ultimate in development, and that each process must be treated in detail as far as improvements are concerned, it is possible to set a lower limit for the energy required for desalination by the use of thermodynamic arguments. This minimum energy requirement applies to all processes, and is analogous to the use of the efficiency of the Carnot cycle for the ultimate limit of power generation economy.

Figure 3-1 represents a generalized desalination plant into which saline water flows at a steady rate and from which there issues a stream of pure water and one of brine. The plant is enclosed in a heat insulating barrier through which energy can be transmitted in the form of work, the latter being either mechanical or electrical, for heat will be treated separately. It is assumed that steady conditions exist, i. e., that flow rates, temperatures, and pressures are constant with time.

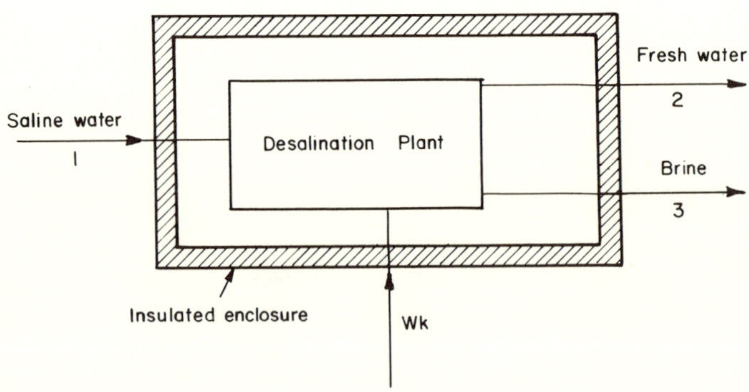

FIG. 3-1. Diagram of generalized desalination plant with mechanical energy input. (Schematic only.)

A. Application Relative to Mechanical or Electrical Energy

Application of the principle of conservation of energy and of the concept of entropy for the conditions in Fig. 3-1 leads to the following equation:

$$-\text{work} = Ts + n_2'(\bar{F}_2' - \bar{F}_3') + n_1'(\bar{F}_3' - \bar{F}_1') + n_2'(n_1''/n_3'')(\bar{F}_3'' - \bar{F}_1'') \quad (9)$$

where work is the energy required to accomplish separation of pure water from the solution, T is the absolute temperature, s is the increase in entropy of the system, n is the number of moles, $\bar{F}$ is the partial molal free energy, 1, 2, and 3 are subscripts: referring to streams in Fig. 3-1, ' is the water component, and " is the salt solution component. Since steady flow conditions are assumed to obtain, Eq. (9) can be written simply as above and the values of work, entropy increase, and the number of moles evaluated in terms of the quantities passing into and out of the plant in some given time. It would be equally correct, but more cumbersome to handle, to express all of the extensive quantities as rates. A derivation of Eq. (9) in terms of quantities has been formulated by Dodge and Eshaya [1], and a derivation in terms of rates by Tribus et al. [2]. Besides steady flow conditions, Eq. (9) assumes no interchange of heat between the desalting plant and the surroundings.

The changes in free energy indicated in Eq. (9) can be expressed in terms of the activity. Making appropriate substitutions for activities, and dividing each term of the equation by n_2', the quantity of pure water produced (in number of moles) results in a relation suitable for calculation, as follows:

$$-\frac{\text{work}}{n_2'} = \frac{Ts}{n_2'} + RT\left[-\ln a_3' + \frac{n_1'}{n_2'}\left(\ln\frac{a_3'}{a_1'} + \frac{n_1''}{n_1'}\ln\frac{a_3''}{a_1''}\right)\right] \quad (10)$$

where a is the activity (dimensionless) and R is the universal gas constant. Note that a_2', the activity of the pure product water, is unity, so that ($\ln a_2'$) is zero and does not appear in the equation. Since it is desired to determine work quantities which are the lowest conceivable, the ideal processes involved will necessarily be reversible, so that the entropy increase is zero and Eq. (10) is reduced to the following:

$$-\frac{\text{work}}{n_2'} = -RT\ln a_3' + RT\frac{n_1'}{n_2'}\left(\ln\frac{a_3'}{a_1'} + \frac{n_1''}{n_1'}\ln\frac{a_3''}{a_1''}\right) \quad (10a)$$

In order to use this equation for calculations, values of the activities of the water and salt components of the solution must be obtained.

While the most direct method to obtain these values would be from curves or tables, these are not always available. Values for the activities for the water component can be estimated from the relation

$$a = p/p_{sat} \tag{11}$$

in which p is the water vapor pressure of the solution and p_{sat} is the saturated water vapor pressure at the given temperature.

This equation assumes that water vapor obeys the perfect gas law, an assumption which is valid at temperatures lower than about 130°F. To apply Eq. (11), values for p can be obtained by determining the boiling point elevation as described in Chapter 2, deducting this value from the desired saturation temperature, and obtaining the reduced pressure by interpolation of steam table values. For example, suppose it is desirable to obtain the activity of normal seawater at 70°F. The boiling point elevation is found to be 0.524°F; therefore, the saturation temperature of the vapor will be less than 70°F by that amount, namely, 70°F - 0.524°F = 69.476°F. By interpolation of steam table values, the saturation pressure corresponding to 69.476°F is found to be 0.3568 psi. The activity of the water component is then the ratio of this pressure to the saturation pressure of vapor at 70°F, or 0.3568/0.3631 = 0.9826. This value is quite close to published figures.

Values for the activities of the salt in seawater solutions are not generally available, but there are published values for the activity of sodium chloride in solution. It is common practice to assume that the activity of the salt in a sea salt solution is equal to that of sodium chloride in a solution of equal molal concentration. When reviewing data on activities it should be borne in mind that the activity of salt in solution is zero at zero concentration and increases almost parabolically with concentration. In contrast, the activity of the water in a solution is unity at zero concentration and decreases almost linearly as salt concentration increases. The numerical value of the activity of the water component never exceeds unity, whereas values for the salt component achieve values greater than one for salt concentrations of 10% or more.

A special case frequently cited in applying Eq. (10a) is that in which the flow rates of water entering and brine leaving are very large compared to the flow rate of the product water. In such a situation, the concentrations of dissolved salts would be very nearly equal in the feedwater stream and the brine stream, so that the last term in Eq. (10a) vanishes with the following result:

$$- \text{work}/n_2' = -RT \ln a_3' \tag{12}$$

Equation (12) expresses the minimum energy required to obtain pure water from a saline solution of a given concentration. Values have been computed from this equation for sea salt solutions of different concentrations at 70°F, and are shown graphically in Fig. 3-2. From this curve it will be noted that normal seawater, 3.44% salt, has a minimum energy of separation of 2.61 kWh per thousand gallons, and that twice-concentrated seawater, with a salinity of 6.88%, has an energy of separation of 5.66 kWh. While the curve of Fig. 3-2 applies strictly to sea salt solutions only, it would be equally valid for any aqueous solution which has the same boiling point elevation as a sea salt solution of equal concentration. The curve may also be used as a first approximation for any aqueous solution containing the same ions as those found in seawater.

Since any practical scheme for separating pure water from a saline solution deals with finite rates of flow in the three streams, i.e., feedwater, pure water, and brine, the feed and brine streams will no longer have equal salinities. Two extreme cases can be described for plants operating with given flow rates — one being the case in which the incoming saline water enters a pool of water from which the separatory device removes pure water, and the other case being one in

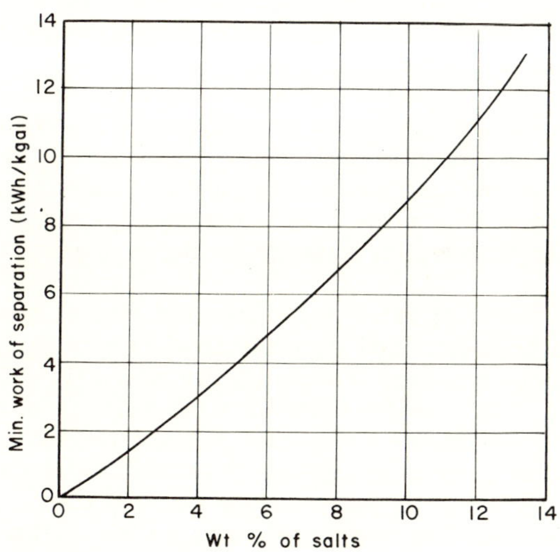

FIG. 3-2. Effect of salinity on minimum work of separation. Recovery of freshwater from seawater and its concentrates. Note: Work of separation based on infinitesimal production of desalted water.

which the saline water flows through a very large number of small devices in series, each one of these devices operating reversibly and contributing a share of the pure water.

In the case of the pool into which the saline water feed discharges, the flow rates of all three streams are assumed to be adjusted so that the level and salinity of the pool remain constant, i.e., the incoming saline water flow equals the sum of the pure water discharged and the brine discarded. The brine pool is assumed to be quite large compared with the rate at which pure water is withdrawn, so that the salinities of the brine pool and the brine pumped out of the pool are equal. Energy of separation values, computed from Eq. (10a), are based upon the activities of the pool and of the reject brine. Since these salinities are equal, the last term of this equation again vanishes, thus permitting the use of the simpler Eq. (12).

To illustrate the magnitudes of the work quantities involved, consider the case in which seawater is added to a pool of brine at a constant rate, and pure water and reject brine are removed from the pool, also at a constant rate. After a sufficient period of time the salinities of the pool and reject brine will become constant and equal, and the power required by the separatory device will also be constant. Its magnitude can be determined from Eq. (12) or from the curve of Fig. 3-2. This calculation was made for a series of different recovery ratios and the results plotted on Fig. 3-3. This case is generally referred to as that of internal mixing, since the saline water is mixed with the water in the system prior to removal of freshwater. The practical device most nearly resembling this idealized case is the simple vapor compression plant, in which seawater is pumped into the evaporator and mixed with the residual brine. Vapor formed from this mixture is then compressed and discharged to the condensing side of the device. Brine rejected from the evaporator has the same salinity as that in the mixture from which water is being evaporated.

The extreme case opposite to the internal mixing concept is one in which pure water is removed from the feed stream in differential amounts as the stream proceeds through the plant. It will be realized that the recovery of pure water increases the salinity of the residual brine. By using a multitude of separate but similar devices through which the feedwater passes in series, it is possible for each device to remove a small amount of water before passing the residual brine on to the next device. The energy required by the first device will be that required for the salinity of the feedwater, while the last device will require the much larger amount of energy for separation, corresponding to the maximum brine concentration. The intermediate devices

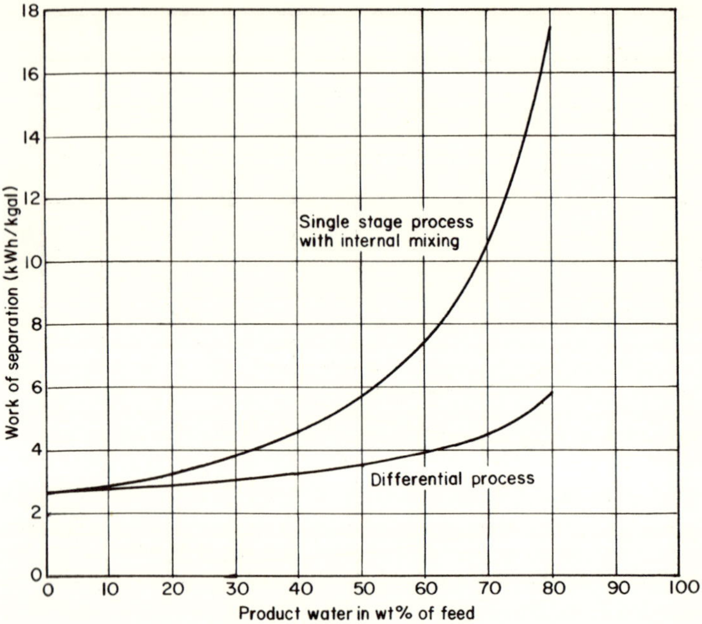

FIG. 3-3. Effect of recovery fraction on work of separation. Recover
of freshwater from seawater.

will require increasing amounts of energy between these two extreme The total energy required by all these devices, which are assumed t be very large in numbers, may be determined from Eq. (10a). Using increasing fractions of pure water recovered from the feedwater in th equation, the results are plotted as the lower curve on Fig. 3-3. Th last term of Eq. (10a) represents the maximum saving due to the avoi ance of internal mixing, and its value is given by the ordinate dista between the two curves at any given fractional recovery of water. From these curves it will be noted that the energy required at 50% recovery of freshwater is 5.66 kWh/kgal for internal mixing, and 3.51 kWh/kgal for the differential process, a saving of energy of 2.15 kWh/kgal. This serves to emphasize the importance of avoiding, or at least minimizing, internal mixing in plant design.

Because the circulation of sizable quantities of water is implie and no energy input is assigned to produce this circulation, Gillila [3] has pointed out that the two curves of Fig. 3-3 can be somewha misleading. For example, the production of freshwater with zero change in salinity between the brine and the incoming seawater wo

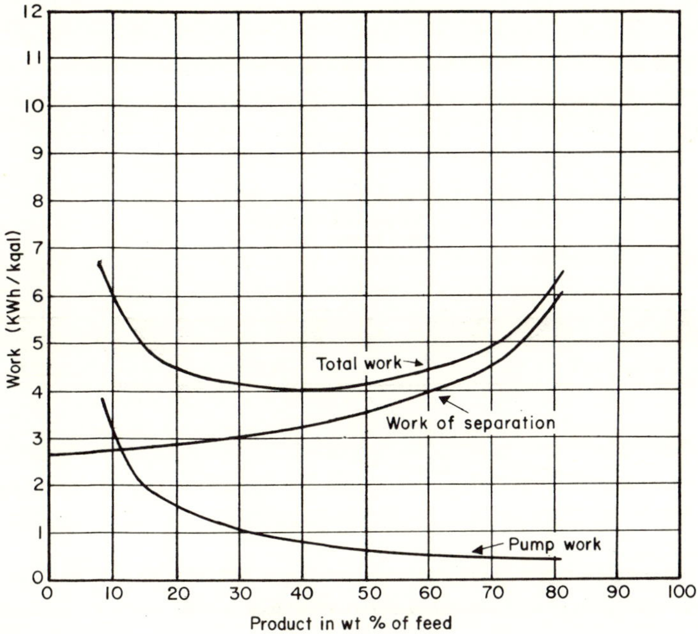

FIG. 3-4. Effect of pump work on total energy requirement. Recovery of freshwater from seawater. Note: Pump work based on 100 ft head and 100% efficient pump. Work of separation based on ideal differential process.

require a very high rate of water flow, theoretically an infinite rate, and this in turn would require a large amount of power for pumping. Since the pumping power would depend upon the head loss through the plant, and there is no particular value which is typical for all desalination processes, some arbitrary quantity must be chosen. Plotted in Fig. 3-4 is a curve for the energy required for pumping, based on ideal pumps delivering water at a head of 100 ft. Also plotted on Fig. 3-4 is the curve for no internal mixing, from Fig. 3-3. By adding the ordinates of these two curves, the third curve of Fig. 3-4 is obtained. It will be noted that this curve has a minimum value near the middle of the plot, indicating that the sum of the energy of separation and the energy required for pumping will be a minimum when about 50% of the seawater is converted into freshwater. Despite the fact that various values of pumping power could shift the minimum point to the right or left, it would not move it far from the 50% point. Significantly, the recovery of about 50% of the seawater input as freshwater is standard

practice in the desalination field, a second reason being related to problems of scale control. Thus, it is appropriate to compare the energy required in an actual plant to the energy of separation at 50% recovery, as given in Fig. 3-3.

B. Application Relative to Heat Energy

The arguments on which the earlier analyses were based included the assumption that the plant would be operated by mechanical energy, represented by the work input of Fig. 3-1. However, the large majority of practical desalination plants use heat energy for the separation process, so that it is essential to determine the minimum heat energy needed to accomplish separation. One approach to determining this value is to conceive a highly efficient heat engine within the boundaries of the plant, and to imagine the mechanical energy produced by this engine to be applied to the above described desalination process. Figure 3-5 shows the desalination plant of Fig. 3-1, modified to include a reversible heat engine operating on the Carnot cycle and supplying mechanical work to the desalination device.

In Fig. 3-5, heat energy is delivered to the system at temperature T_1 and rejected at temperature T_2, so that the engine does work amounting to

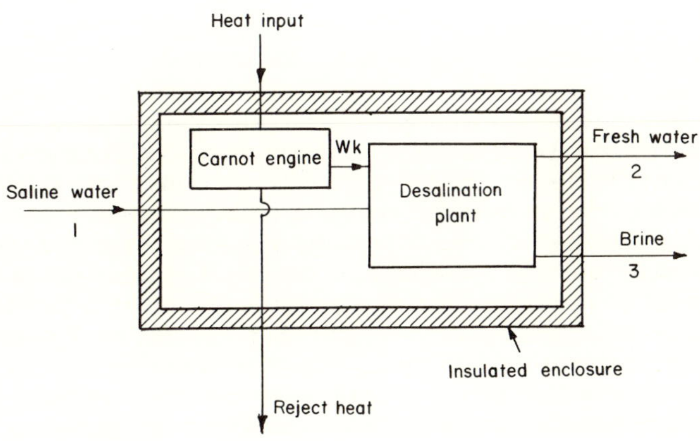

FIG. 3-5. Diagram of generalized desalination plant with heat energy input. (Schematic only.)

$$\text{work} = Q(T_1 - T_2)/T_1 \tag{13}$$

in which Q is the heat input and $(T_1 - T_2)/T_1$ is the Carnot efficiency.

Thus, the heat input required for the separation process will equal the mechanical work input, determined by Eqs. (10a) or (12), divided by the appropriate Carnot efficiency. For example, many distillation plants receive heat by condensing steam at 250°F and reject heat in the cooling water at 70°F. The Carnot efficiency of an engine operating between these temperatures is 25.4%. Using the reciprocal of this fraction gives a factor of 3.93, which means that the minimum heat input to a plant using heat energy under the given temperature conditions would be 3.93 times the quantities read from the curves of Figs. 3-2, 3-3, and 3-4. For instance, from Fig. 3-3, the recovery of freshwater from seawater by a differential process would require a heat input of 14.1 kWh/kgal at 50% recovery of freshwater. This is equivalent to 5.79 Btu/lb of freshwater, which, when divided into the latent heat of steam at 250°F, would indicate a theoretical gained output ratio of 163 lb of freshwater/lb of steam. For comparison it may be noted that the most efficient distillation plants now in operation show actual gained output ratios of about one-tenth the above, namely, 15 to 20, with the majority of the plants not over 15. Research and development efforts to bring the actual gained output ratios closer to the theoretical values will be dealt with in the several chapters devoted to specific processes.

III. HEAT REJECTION PROBLEMS

The previous paragraphs have dealt with the minimum energy required by ideal reversible desalination processes. This is the energy that is just sufficient to do the work of separating the freshwater from the saline solution isothermally. For processes which are not ideal, more energy will be required by the desalination plant and the excess of energy used must be rejected from the plant if temperatures within the plant are to remain constant. For example, an efficient practical vapor compression plant will require about 45 kWh/kgal of water produced, and operates with internal mixing at about 50% recovery of freshwater. From Fig. 3-2 it is noted that about 4 kWh/kgal will be needed to accomplish the reversible separation of freshwater from brine. The balance of the input, 41 kWh/kgal, must be rejected from the plant. If all of this energy is rejected as heat in the outgoing brine and product streams, a temperature elevation above the inlet seawater temperature of about 8.5 °F would be sufficient to carry away the heat. This temperature rise would be quite acceptable for both reject brine and product.

The heat which must be rejected from plants which use heat energy, as such, is somewhat larger and may be approximated by assuming the vapor compression plant noted above to be driven by an internal combustion engine. The heat to be rejected by the engine will be approximately twice the heat equivalent of its output, thus requiring the disposal of an additional 90 kWh/kgal of freshwater produced. If it were possible to use the exiting streams of brine and cooling water to collect this heat from the engine, the rise in temperature of these streams would be nearly 19 °F above the 8.5 °F rise noted above. Since this total temperature elevation of 27.5 °F above the incoming seawater would be quite unacceptable in either the brine stream or the product stream, it is usual to have an additional stream of cooling water sufficiently large in volume so as to limit its temperature rise to some predetermined low value, the magnitude of which depends upon thermal pollution considerations. It will be seen in later chapters that distillation plants using heat energy reject somewhat more heat than the combination of engine and vapor compression device noted in the above example.

IV. DISPOSAL OF WASTE BRINE AND COOLING WATER

The discharge of waste brine and cooling water from a desalination plant must be accomplished with zero, or at least minimum, harmful change to the local aquatic environment. If the plant in question is a distillation facility, the combined cooling water and waste brine streams are likely to be 12-15 °F warmer than the inlet saline water. They will also contain all of the dissolved materials removed from the product water, plus a probable 0.15 ppm of copper from the unavoidable corrosion of the copper alloy tubing used in the heat exchangers. It is feared that these wastes are unfavorable and even toxic to the marine organisms which constitute the basic food chain in the receiving waters. Natural evolution has established a marine ecology of diatoms, phytoplanktons, dinoflagellates, and many different varieties of algae, etc., which grow most favorably in waters having a narrow range of temperature and salinity, the specific ranges differing with the particular strain or species involved. Any major disturbance, such as that which would be produced by the discharge of warm salty water from a newly established desalination plant, would thus likely seriously interrupt or partially destroy the ecological food production in the receiving waters.

The effect of copper ions and certain other particles in the discharge water can be quite serious, since these materials tend to accumulate in many organisms rather than being passed out with other wastes. In the open ocean the concentration of copper is very low, about 0.003 ppm, and organisms there do not seem to be affected by this concentration. However, in localities where concentrations from industrial waste are even as little as 0.05 ppm, it has been observed that this small amount has a toxic effect on some species of marine life. It has been recommended that the maximum concentration of copper should not be more than 0.02 ppm. Reduction of the copper content of any discharge may be accomplished by dilution, or by removal through ion exchange or some other chemical treatment. The latter two have been experimented with and, although they have been found to be effective, the cost of either process has been found to be approximately $0.04 per kgal of distillate, thus adding an almost prohibitively high element to the total desalination cost. For this reason, dilution is resorted to as the most acceptable solution.

Dilution of the effluent streams by means of jets diffuses the streams of brine and cooling water, and thereby reduces the temperature, salinity, and copper concentration by rapid mixing with the receiving water. Plants located on the shoreline can effectively utilize jets of this type to disperse the waste brine and cooling water into the ocean. An extensive study of such dispersion was made by LeGros et al. [4], based on model experiments as well as on a collection of data from existing plants. In general, this study showed that when dispersion was effective in reducing dissolved copper concentration to the desired 0.02 ppm, a satisfactory reduction of temperature and salinity was accomplished at the same time.

For plants located at inland sites, there is generally no body of water large enough to accept the cooling water and waste brine without objectionable alteration of temperature and composition in the receiving pond. In such cases the designer has open to him essentially four choices — namely, evaporation in open ponds, discharge to deep underground porous strata by deep bore holes, transport through closed pipelines to the ocean for dispersal there, and utilization of the waste streams for the production of salable chemicals.

A review of these several alternatives was prepared by DePuy [5] and seems to favor evaporation in open ponds. Underground disposal of waste brines is rather extensively used by the petroleum industry but is not favored for desalting plants because of the relatively much larger quantities of brine involved. Pipeline transportation of brines to the ocean is rather costly and would be prohibitively expensive for plants located more than a very few miles from the coastlines. The quantity of common chemicals which could be recovered from desalting plant brines would greatly exceed the demand for such chemicals. Thus, it appears that solar evaporation in open ponds is the most realistic scheme.

Considerable statistics are available concerning solar evaporation ponds, which have long been used rather extensively for the production of common salt, i.e., sodium chloride, from seawater. Such ponds are generally located adjacent to bays or other seawater inlets, as, for example, the salt ponds located on the southeast side of San Francisco Bay. At this location weather records show an annual evaporation of some 30-50 in., chiefly in the summer and autumn months, and an annual rainfall of around 25 in., chiefly between December and March. Here, saltmaking begins about the first of May, and the last of the salt is harvested in November.

If inland ponds of this type were to be used for the disposal of brine from desalting plants, it is clear that the bottom of the pond must be sealed to prevent leakage into underground strata, that it must be located at a place where evaporation exceeds rainfall by a substantial amount, preferably about 20 in., as suggested by DePuy. This represents an annual net evaporation of about 12.5 gallons per ft^2. For a plant producing one million gpd pure water with a 50% recovery factor, the pond area required to evaporate one million gpd brine disposal would amount to about 700 acres.

As can be readily understood from the foregoing, brine disposal is a major problem in the engineering design of any desalting facility, whether the plant is to be located on the seacoast or at an inland point. Solution of this problem will involve both capital and operating expenses which will be a significant addition to the cost of the water produced.

V. SUMMARY

The objective of the engineering design of a desalting facility is to produce freshwater at a minimum cost. Recent emphasis on quality of the environment makes it an added mandate that the disposal of waste heat and brine be accomplished without damage to the surroundings or to life processes therein.

Current design methods, coupled with the use of digital computers, enable the designer to optimize the plant from the viewpoint of water cost. It can be shown that the minimum water cost will be achieved when the cost of energy to operate the plant is approximately equal to the cost of capital amortization. This narrows the range of the variables which must be explored in searching for the optimum design.

There is a theoretical minimum energy required for separating pure water from saline water. This minimum value varies with the salinity of the water and with the fraction of the saline water which is recovered as pure water. If the energy required for pumping water through the plant is added to the energy of separation, the resulting sum will be a minimum when about one-half of the feedwater appears as freshwater, or in other words, the plant shows a 50% recovery. This minimum energy, when calculated on the basis of mechanical energy required for seawater plants, is about 4 kWh/kgal. A comparison with actual plants shows a wide divergence between this hypothetical minimum and the actual energy requirement. Research and development efforts aim to make possible more efficient plants, requiring less energy.

The disposal of waste heat and concentrated brine in a manner which does not alter organic life in the immediate surroundings is a major problem for the designer. For plants located on the seashore the accepted solution seems to be dispersal into the ocean by jets so arranged that the mixture of effluent and entrained ocean water is so dilute that no damage is done to the biota in the area. For plants at inland locations the favored method is that of solar evaporation from large ponds.

REFERENCES

1. B.F. Dodge and A.M. Eshaya, Thermodynamics of Some Desalting Processes, Saline Water Conversion, No. 27, Advances in Chemistry Series, American Chemical Society, 1960, pp. 7-20.
2. M. Tribus et al., Thermodynamic and Economic Considerations in the Preparation of Fresh Water from the Sea, Rept. No. 59-34, Dept. of Engineering, Univ. of Calif., Los Angeles, June, 1959.

3. E. R. Gilliland, Fresh Water for the Future, Ind. Eng. Chem., 47 (No. 12), Dec. 1955, pp. 2410-2422.
4. P. G. LeGros et al., A Study of the Disposal of the Effluent from a Large Desalination Plant, Research and Development Progr. Rept. No. 316, Office of Saline Water, U. S. Dept. Interior, Jan. 1968, 491 pp.
5. G. W. DePuy, Disposal of Brine Effluents from Inland Desalting Plants: Review and Bibliography, Research and Development Progr Rept. No. 454, Office of Saline Water, U. S. Dept. Interior, July 1969, 211 pp.

Chapter 4

METHODS OF DESALINATION

I. VARIETY OF CLASSIFICATIONS

Numerous processes have been proposed for the desalination of water, and a few of these have attained important commercial status. Distillation, electrodialysis, and reverse osmosis are already applied to a variety of practical installations. Solvent extraction, ion exchange, freeze separation, and hydrate separation have been less developed, and are limited to special cases. Of the methods that are of current importance, electrodialysis and reverse osmosis are best adapted to the desalination of brackish waters while distillation is applicable to the entire range of salinities up to seawater as dense as 43, 000 ppm. This chapter will undertake to give the essential features of several processes. Each particular desalination method will be discussed in detail in later chapters.

It may be noted that the several processes for desalination in this chapter have been grouped according to the phenomena involved, e. g., those involving phase change in water, as in distillation, freeze separation, and hydrate separation; those utilizing surface properties of membranes, as in electrodialysis and reverse osmosis; and those utilizing ion-selective properties of solids and liquids, as in ion exchange and solvent extraction. Another classification could be one which would involve the various forms of energy, namely, heat,

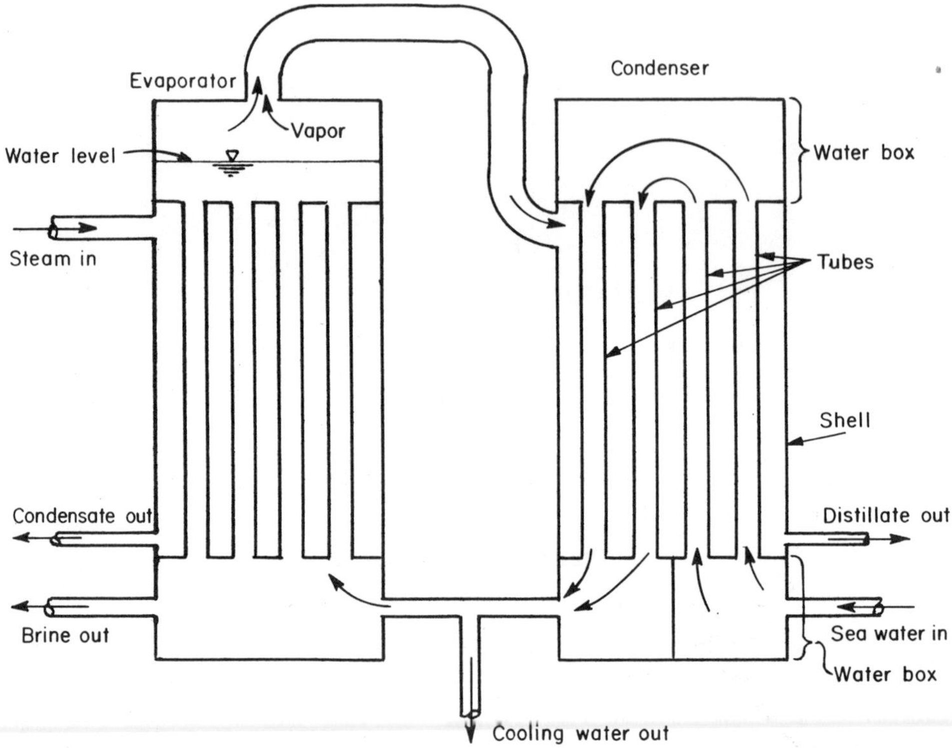

FIG. 4-1. Schematic cross section of simple distiller.

mechanical, electrical or chemical. The usefulness of this latter basis of grouping would have bearing upon the relative energy costs of the different processes, since heat energy is the cheapest while chemical energy is the most expensive. Distillation may be accomplished using either heat or mechanical energy, reverse osmosis requires mechanical energy, electrodialysis requires electrical energy, and ion exchange requires chemical energy. A further classification is one which distinquishes between the withdrawal of water from solution, and the extraction of salts from solution, as is exemplified by the comparison of any distillation process with electrodialysis. Distillation involves the withdrawal of water vapor; electrodialysis requires the expulsion of ions from solution. Useful as these latter classifications may be, the grouping according to the phenomenon involved is regarded as more significant than the others from the viewpoint of the nature of water, and hence is the basis for the organization of this chapter.

II. PROCESSES USING A PHASE CHANGE OF WATER

A. Distillation

Distillation, as a process for desalination, is effective because most of the chemical materials found in saline waters are nonvolatile at the temperatures normally employed, and hence remain in the unevaporated brine. It should be noted at this point that the term "distillation" used in this book, and generally accepted in the field of desalination, does not have the same meaning that is normally reserved by chemical engineers for processes involving the separation of two or more volatile liquids which are mutually soluble. The term "evaporation" is used by them to indicate the process of separating pure water from aqueous solutions of nonvolatile materials. Although there have been sporadic attempts to introduce the use of the word "evaporation" to indicate the production of freshwater from saline sources, the word "distillation" has become firmly established in the technical press, and will be used throughout this book.

1. Simple Distillation

The distillation process involves three discrete steps to accomplish its purpose:

1. Vapor formation due to the addition of heat to a body of saline water

2. Removal of this vapor from contact with the liquid from which it came

3. Condensation of the vapor due to the removal of heat, usually by contact between the vapor and a cooled surface

Figure 4-1 is a schematic diagram of a simple distillation process for recovering freshwater from seawater, and illustrates a practical means for effecting these above three steps. It shows two chambers, an evaporator in which vapor is formed from heated seawater, and a condenser to condense the vapor to distillate. Each chamber consists of a tube bundle enclosed in a shell, with water boxes at the ends of the tube bundles to permit the conduction of fluids to and from the tubes. Vapor is formed in the tubes of the evaporator, because of condensation of steam from an external source on the outside walls of

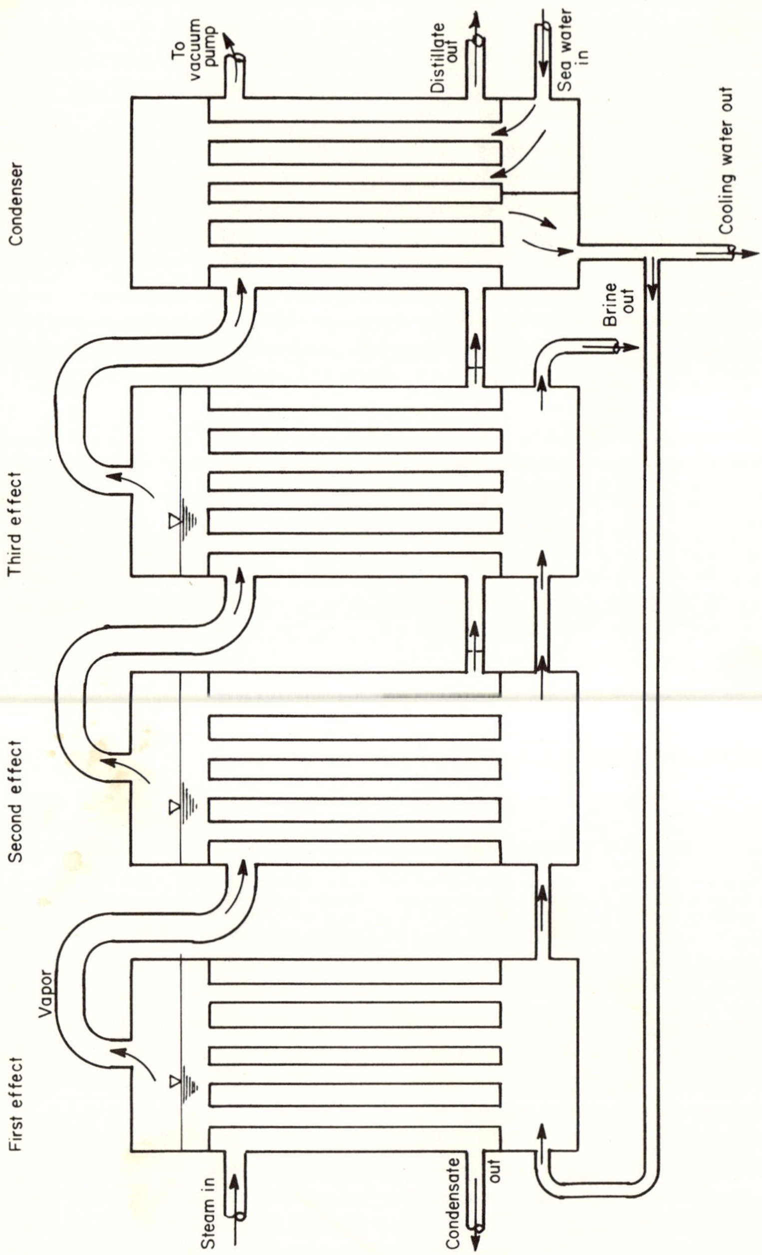

FIG. 4-2. Schematic cross section of a triple-effect distillation plant.

the tubes, generally referred to as the shell side of the tubes. The vapor leaves the evaporator and enters the shell side of the condenser, where it is liquefied by contact with the cold surface of the condenser tubes, and is discharged as distillate. Cold seawater flows through the condenser tubes to absorb heat from the condensing vapor, and is heated thereby. Leaving the condenser, this slightly warmed seawater is split into two streams, a portion of it going to the tubes of the evaporator and the balance being discarded for heat rejection. It should be noted that the water discarded for heat rejection must be sufficient in quantity and high enough in temperature above the inlet value to carry away all of the heat which was added by the condensation of the external steam. The water flowing to the evaporator must be equal in quantity to the amount of the vapor produced plus the brine which is blowdown. The rate of brine blowdown must be such as to prevent the accumulation of dissolved salts above a predetermined value.

In order for the device shown in Fig. 4-1 to function, and for heat to be transferred from steam condensing on the outside of the tubes in the evaporator to the water on the inside, the temperature of the steam on the outside must be higher than the boiling temperature of the water within the tubes. Due to the relation between the boiling temperature and pressure of water it will therefore follow that the pressure within the tubes will be less than on the shell side, and that the latent heat of the water therein will be greater than for the higher pressure and temperature steam on the outside. As a consequence of this latent heat differential, the condensation of 1 lb of steam input will result in the production of less than 1 lb of vapor in the simple distillation device described.

2. Multiple-Effect Distillation

It has been found that is possible to multiply the amount of vapor produced by a given amount of initial steam by placing several evaporators in series, with the pressures of the evaporating sides being maintained at successively lower values to ensure heat flow. Figure 4-2 shows the essential features of a triple-effect distillation plant containing three evaporators. The first evaporator, like that in Fig. 4-1, receives its heat from a separate steam supply, the only condition imposed being that the steam in the shell side must be higher in temperature than the boiling temperature of the brine in the tubes of the evaporator. The vapor produced in the tubes of the first evaporator is condensed on the shell side of the tubes in the second evaporator, thereby providing heat for the evaporation of some of the water which is inside of the tubes. To cause this flow of heat from the condensing steam to the water inside the tubes, it is necessary that the water temperature be less than the condensing steam temperature. The vapor formed will be at the same temperature as the water, so that the vapor in this second effect will have a temperature and pressure less than that which was formed in the first effect. The vapor leaving the second evaporator is condensed on the shell side of the tubes in the third evaporator, producing in the tubes a third increment of vapor which is condensed on the shell side of the condenser. Inasmuch as each pound of steam or vapor condensed in the shell results in the formation within the tubes of just slightly less than a pound of vapor, it is clear that the condensation of 1 lb of steam in the first evaporator will result in the production of a little less than 3 lb of water by the time the last vapor is liquefied in the condenser. This multiplication of the effect of condensing steam in the successive evaporators leads to the designation of this scheme by the term "multiple-effect" distillation.

The schematic diagram of Fig. 4-2 does not show all of the auxiliary equipment required to operate this type of distiller. A vacuum pump is indicated as being connected to the condenser, and has the two functions of removing air from the system and producing a pressure gradient through the plant so that brine and vapor will move from one effect to the next. In an actual plant there would likely be vacuum pump connections to one or more of the effects where air might be released from the brine. Also because of the vacuum, pumps would be required for removing the brine and distillate from the system. This information is included in detail in later chapters, so that the diagram of Fig. 4-2 should be regarded as limited to its stated purpose of illustrating the multiple-effect concept.

One of the indices used to specify the steam economy of multiple-effect plants is the "gained output ratio" (GOR), and is equal to the amount of distillate produced divided by the amount of heating steam used in the high-pressure evaporator. Numerically the GOR will be slightly less than the number of effects for several reasons. Some of the heat supplied by the condensing steam is absorbed by the heating of cold feedwater from the low inlet temperature to the boiling temperature, and some of it is lost to the surroundings due to less than perfect insulation. A less obvious cause for the decrease of the GOR to a value less than the number of effects is the fact that the latent heat increases as the pressure decreases, so that the amount of vapor formed in each successive effect will be less than that formed in the previous effect.

While the GOR for a given number of effects will vary slightly from one plant to another, a first approximation to its value can be made by multiplying the number of effects by 0.8. Thus it would be expected that a plant with six effects would show a GOR of about 4.8.

The GOR, as defined above, is a true measure of economy for plants in which pumps and other machinery are steam-driven, where all of the steam needed by the engines which drive the machinery is delivered to the brine heater of the distiller plant after expansion in the engines or turbines. For plants which contain electrically driven machinery, receiving power from a source external to the distillation plant, the definition of the GOR must be modified to include such inputs, if it is to serve as a true measure of economy. Schemes for modifying the GOR for this situation are discussed in a later chapter.

The steam economy may also be expressed in terms of the "performance ratio" (PR), which is defined as the number of pounds of distillate produced per 1000 Btu of heat input. The PR is preferred by some writers for the comparison of distiller plants because it is independent of the external steam conditions, whereas the magnitude of the GOR will vary with the pressure at which external steam is furnished. In general, the PR will be slightly larger than the GOR, since heating steam is usually provided at pressures close to atmospheric, at which condition the latent heat is less than 1000 Btu per pound. For these reasons, although the GOR and the PR differ only slightly, numerically, it is important to ascertain which one has been used in any specific report in order to adequately compare performance. Further comments concerning these relationships in multiple-effect distillation will be dealt with in Chapter 6.

3. Flash Distillation

Another method of effecting economy in distillation is by means of flash distillation plants. "Flash" evaporation is the production of vapor by the sudden release of pressure on a hot liquid, which is near its boiling temperature initially. When the pressure is reduced, the boiling temperature is correspondingly decreased, requiring a reduction of the sensible heat in the liquid and thereby furnishing heat for the production of a small amount of vapor. Flash distillation plants, using this scheme for producing vapor, are designed so that the incoming feedwater is heated through its entire range of temperature under pressure conditions which do not permit the formation of vapor during the heating process. After the maximum desired temperature has been reached, the pressure on the liquid is reduced in a number of stages. At each step in the reduction of pressure there is an accompanying formation of vapor by flashing. The condensation of all of these increments of water constitutes the water output of the plant.

Figure 4-3 shows the essential features of a multiple-stage flash plant. The first stage and the last two stages are shown in the figure,

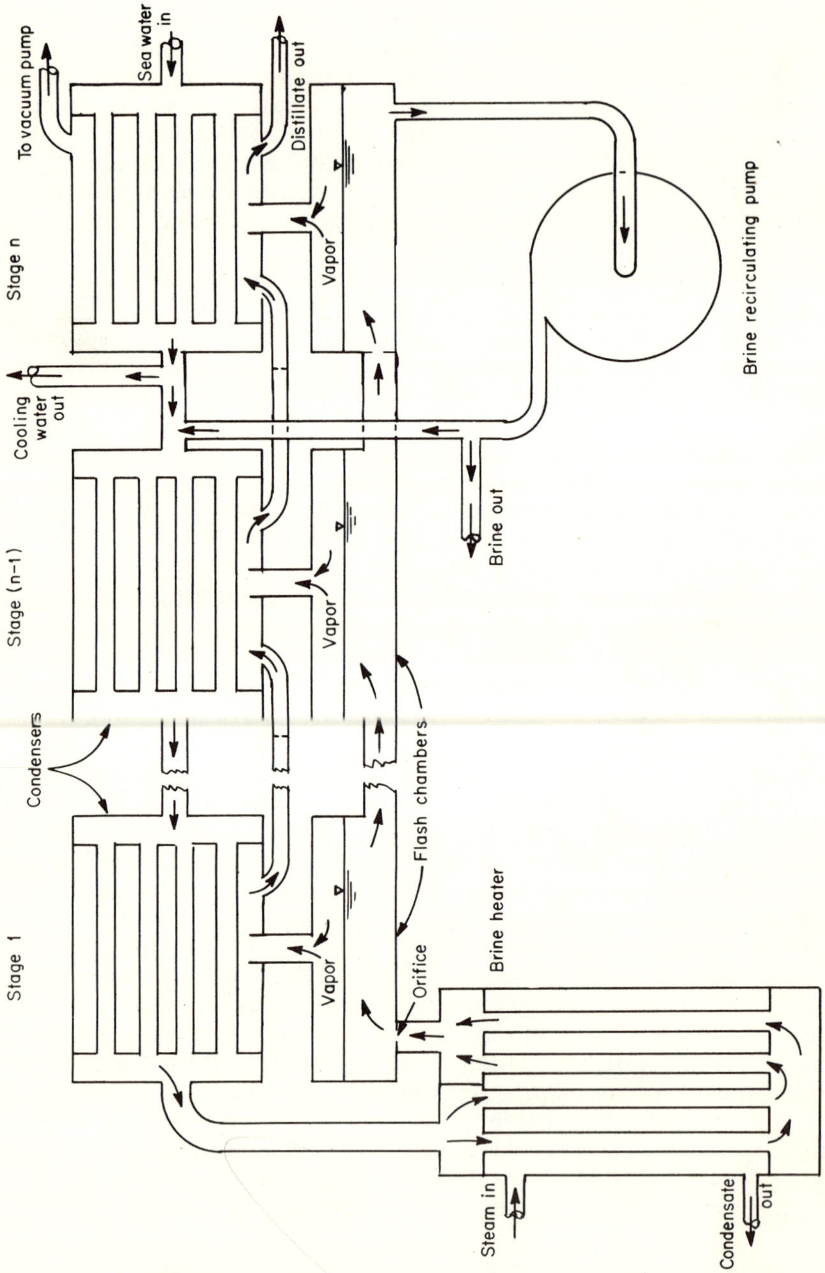

FIG. 4-3. Multiple-stage flash distillation plant. (Schematic cross section.)

with a break indicated where a number of stages similar to stage 1 could be inserted. Each stage is seen to consist of a flash chamber below a bank of condenser tubes. Seawater enters the condenser tubes of stage n, and is heated by condensation of the vapor rising from the flash chamber of this stage. Leaving this stage, the slightly warmed water is divided into two parts, much of it being discharged as cooling water. The remainder of the water is mixed with the warm brine from the recirculating pump to conserve heat, and passed successively through the other stages. Finally, this warmed water is heated to its maximum temperature by passing through the tubes of the brine heater, the heat being furnished by the condensation of steam from an external source.

Having achieved its highest temperature, the hot brine is expanded in a series of small steps to its final low temperature. It first enters stage 1 through an orifice at the entrance to the stage, which serves to reduce the pressure below the saturation value corresponding to the temperature upstream from the orifice. As already explained, this pressure decrease results in the cooling of the stream of brine and the simultaneous production of a small quantity of vapor. This vapor rises in the shell and condenses on the relatively cool outer surface of the tubes through which the incoming water is flowing. The condensate drips off the tubes, is collected in the bottom of the condenser shell, and is passed through an orifice into the condenser shell of stage 2. The unevaporated portion of the hot brine at the bottom of the flash chamber of stage 1 flows through an orifice into the flash chamber of stage 2, where the processes of vapor formation and condensation are repeated. The warm brine and condensate flow separately through all the remaining stages in series, with the stream of condensate increasing at each stage and that of the brine decreasing. The distilled condensate is removed from the plant at the outlet of stage n. The brine is pumped out of this stage, part of it being discharged and the remainder being returned to stage n-1 for recirculation.

As indicated above, the expansion of brine from one stage to the next is accompanied by a decrease in temperature. In designing a plant it is conceivable that either the pressure differentials between stages or the temperature differentials could be made equal. From a practical viewpoint the latter course is generally chosen since equal temperature drops result in approximately equal amounts of vapor produced and require nearly the same amount of heat-transfer surface per stage. If equal pressure decreases were used, the amount of vapor produced per stage would vary greatly, and the area required for heat transfer would need to be varied accordingly. The use of equal heat-transfer areas in the several stages simplifies the design and construction of the distiller units, and hence is the preferred practice.

Most production plants using the multiple-stage flash scheme contain from 25 to 40 stages, spanning a total temperature interval of approximately 100°F. For example, a 40-stage plant would have a per stage temperature drop of 2.5°F. By this means, condensing vapor is made available at a series of temperatures which in turn serves to heat the incoming brine through a large range of total temperature. Figure 4-3 implies a separate casing for each stage, but in actual practice several stages are combined in a single casing. Arranged horizontally, the casing is divided into stages by transverse partitions. Brine is passed from stage to stage through orifices near the bottom of the partitions. The condenser tubes may run from end to end of the casing, with tight-fitting holes or packing where they pierce the several stage partitions, or they may be run perpendicular to the length of the casing, so that each stage has a separate set of tubes. The latter arrangement is favored by manufacturers in the United Kingdom, while the longitudinal arrangement is used almost exclusively by those in the United States.

a. Contrast with Multiple Effect: Quantity of Feed. Flash distillation plants differ from multiple-effect distillers in two important respects: (1) the amount of water flowing through a flash plant is many times greater than the distillate formed and (2) the value of the GOR is relatively independent of the number of stages. To understand the reason for the excessive amount of feedwater needed, it should be recognized that "flash" evaporation is the familiar throttling process, in which flow through an orifice or other restriction is characterized by the equality of the thermodynamic property, enthalpy, upstream and downstream from the orifice. This equality may be expressed in the form of an equation as follows:

$$\text{enthalpy} = h_1 = h_2 \tag{1}$$

in which subscripts 1 and 2 refer to upstream and downstream, respectively. If the upstream condition is that of saturated water, then Eq. (1) can be rewritten as

$$h_{f1} = h_{f2} + x_2 h_{fg2} \tag{2}$$

where h_f is enthalpy of saturated water, h_{fg} is latent heat of evaporation, and x is "quality" of water vapor. For example, consider the range of temperature which has been used in some distillation plants where water is heated to 200° and discharged at 80°F. Obtaining the values of enthalpy from the Steam Tables and substituting in Eq. (2) gives the result for the quality of water vapor as follows:

$$167.99 = 48.02 + 1048.6x$$
$$x = 11.43\%$$

This means that for the temperature span of the above example only 11.43% of the water would be evaporated by the total expansion. The formation of this small percentage of vapor by flashing is the reason why large quantities of water must be circulated in a flash distillation plant. Using the temperature range noted above it would be necessary to expand about 8.75 lb of seawater in order to produce 1 lb of vapor. For comparison, the amount of seawater entering the first effect in a multiple-effect plant is usually designed to be only about twice the expected total amount of distillate from the plant.

Having determined that a very large amount of water must be circulated through the flash distillation plant, the question arises as to whether this large quantity of water is to be pumped from the sea, used on a "once-through" basis and returned to the sea, or whether it can be recirculated to advantage. Water recirculation seems to show significant economies as compared to the "once-through" operation, because the warm brine saves the heat energy that would otherwise be required to warm the cool incoming seawater to the desired brine temperature. Furthermore, seawater must be treated to prevent scale formation before it can be circulated through the high temperature end of the plant. The cost of this treatment increases with the amount of water pumped from the sea; hence, the desirability of reusing the treated seawater. From the standpoint of scale control, the salinity of the recirculated brine may be allowed to increase to about twice the value of seawater without serious effect. Using the temperature limits suggested in the previous paragraphs, with 8.75 lb of brine circulated for each pound of vapor produced, the recirculation of 6.75 lb of brine remaining after evaporation would result in a final salinity of the brine at the end of expansion of twice that of the incoming seawater. Recirculation is practiced in nearly all modern flash distillation plants. This is in contrast to multiple-effect plants in which recirculation is not utilized because of the smaller amounts of feedwater needed to produce the same amount of vapor.

b. Contrast with Multiple Effect: Number of Stages. In order for flash distillation to be economical in its production of vapor, it has been found that it is necessary to cause flash expansion to occur in many steps. The condensation of the many increments of vapor formed in a series of decreasing temperature steps becomes the source of heat for warming the incoming feedwater. The higher the temperature that this incoming feedwater can achieve through this step-by-step condensation of the flashing vapor, the smaller the amount of external heat which must be added to the feedwater, and hence the higher the steam economy.

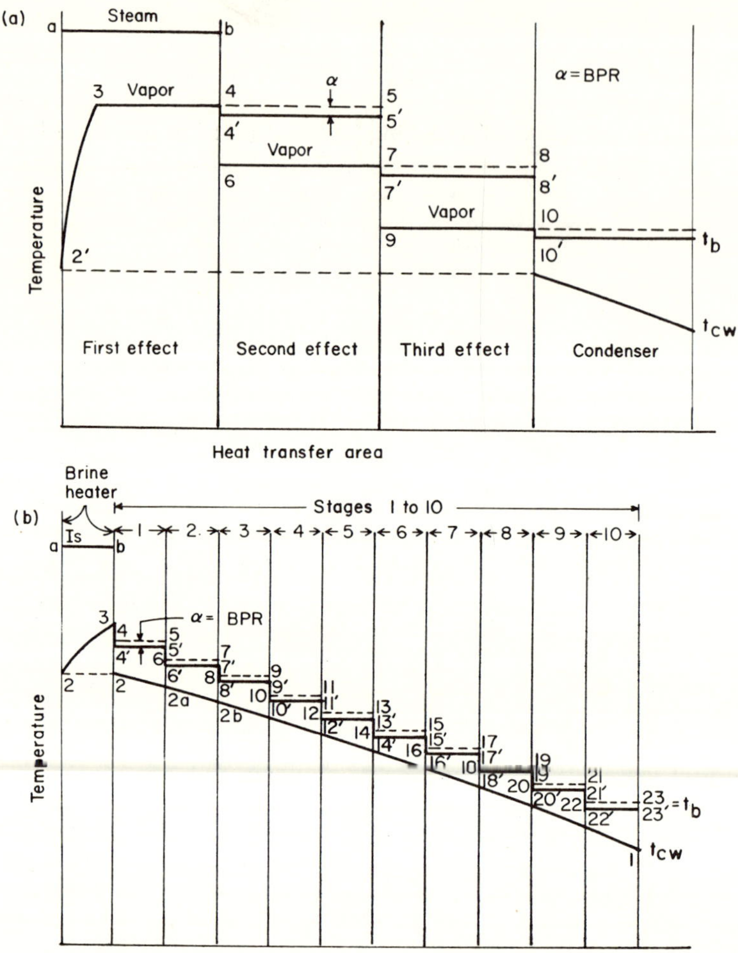

FIG. 4-4. Diagram of temperature profiles of a multiple-effect (a) and a multiple-stage flash distillation plant (b).

Although the economy of multiple-stage flash distillation is enhanced by a greater number of steps, or stages, the relationship is not a direct proportional one, as in the case of multiple effect. This fact is best exhibited by comparison of the temperature profiles of multiple-effect and multiple-stage flash plants. These are shown in Fig. 4-4, with (a) showing the profile of the multiple-effect plant and (b) showing that of the multiple-stage flash plant. In these profiles horizontal distances are proportional to the heat-transfer areas of the plants, and vertical distances are proportional to temperatures.

In Fig. 4-4a, relating to the multiple-effect plant, the line segment a-b shows the temperature at which external steam is condensed on the outside of the tubes in the first effect, and its length is proportional to the area of the tube walls. The line segment 3-4 shows the temperature at which vapor is formed on the inside of the same tubes. The vertical distance b-4 gives the temperature difference which causes heat flow between condensing steam and evaporating liquid. Vapor formed in the first effect passes into the second effect and condenses along the line segment 4'-5', the drop in temperature from 4-4' representing the boiling point rise in the second effect. The unevaporated water in the first effect passes through an orifice into the second effect and evaporates along the line 6-7. Similar processes occur in each successive effect, producing the stepped curve shown in the figure. At each step the upper solid line shows the condensation of vapor and the lower line shows the evaporation process. Since each step produces approximately an equal increment of vapor, and hence of condensate from this vapor, it will be clear that the GOR will be proportional to the number of steps, i.e., to the number of effects.

In Fig. 4-4b, line segment a-b shows the temperature at which external steam is condensed in the brine heater, and the line 2-3 shows the temperature change in the brine which occurs simultaneously. The line 3-4 represents the temperature drop during the flashing process as the heated brine passes through the orifice into the first stage. The boiling point rise of the brine here results in the temperature drop in the vapor from 4 to 4', so that condensation of the vapor takes place along the line 4'-5'. Liquid in the tubes of this stage is heated from 2a to 2, so that the brine enters the brine heater at temperature 2. The unevaporated brine, leaving the first stage, expands through an orifice into the second stage, with its expansion temperature drop represented by the line 5-6. Vapor produced thereby is condensed in the second stage along the line 6'-7', while the liquid in the tubes of this stage is heated from 2b to 2a. The processes of expansion, evaporation, and condensation are repeated in each stage until the brine leaves the plant at point 23, and the condensate leaves at point 23' in the 10-stage plant depicted. From an examination of this figure it will be clear that the amount of heat added by the external steam must be just sufficient to raise the temperature of water from temperature 2 to temperature 3. The number of steps on the curve could be increased or decreased, provided only that the temperature 4' is sufficiently above temperature 2 to produce the required heat transfer. Thus it can be deduced from the diagram that there is some minimum number of stages which would result in a situation in which 4' would be so close to temperature 2 that transfer of heat would be impaired and thereby would require a maximum heat transfer area. At the other extreme, a very large number of steps on a curve with fixed points 2, 3, and 22

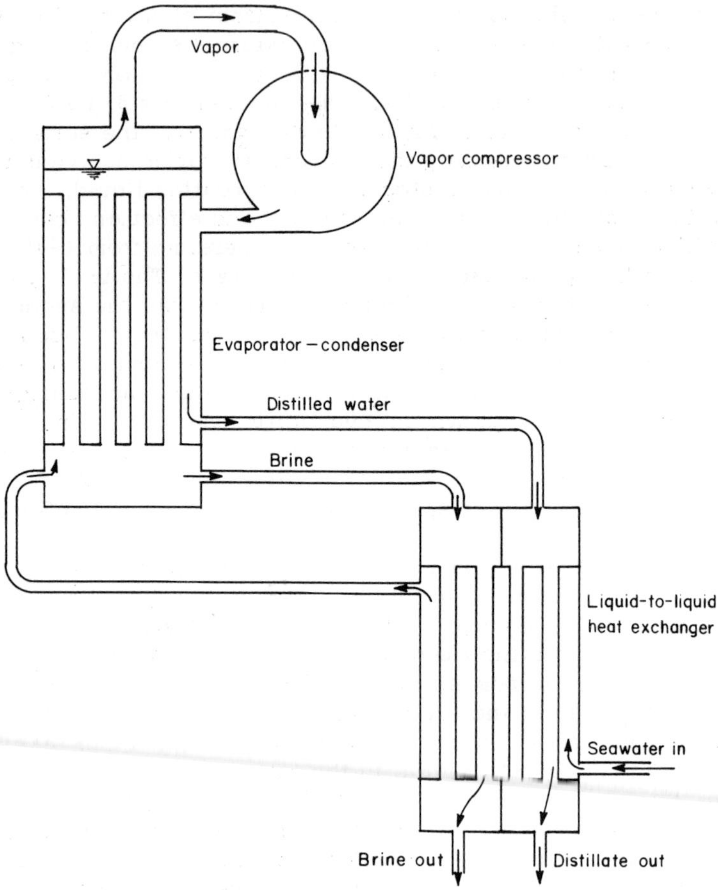

FIG. 4-5. Vapor compression distillation plant. (Schematic cross section.)

would create sufficiently great differences between the temperature of evaporation and the corresponding temperature of the warming brine in each stage, that it would result in the need for a minimum heat-transfer area.

The relationship between the temperatures in the plant and the GOR can be readily formulated. The amount of vapor formed and condensed will be proportional to the temperature drop, t_3-t_{22}, the total flashing range of the brine. The amount of heat added by the external steam will be proportional to the temperature rise, t_3-t_2. Since the

GOR is the ratio of the water produced to the weight of the external steam used, it can be expressed as:

$$GOR = C(t_3-t_{22})/(t_3-t_2)$$

Since C is a constant for any set of values for t_2, t_3, and t_{22}, the GOR will have a single fixed value regardless of the number of stages, i.e., the number of steps on the curve between t_3 and t_{22}. Further discussion of this method may be found in Chapter 7.

4. Vapor Compression Distillation

The distillation processes already described are based upon the stepwise degradation of heat energy from some elevated initial temperature to the final temperature of the cooling water. In contrast, vapor compression distillation is a process in which a mechanically driven heat pump is employed to recycle the latent heat from the evaporating side of the heat-transfer surface to the condensing side of the same surface. This process is shown schematically in Fig. 4-5. In this process the vapor produced is compressed to a pressure sufficiently high that its condensation will supply heat for the production of additional vapor from the same source of water. In the figure it will be seen that the three basic steps of distillation are provided, namely, evaporation, separation of the vapor from the liquid, and condensation. A supplementary heat exchanger is also provided to further cool the distillate and the brine, and to preheat the incoming seawater.

For steady state conditions the heat which must be rejected from the system is that equivalent to the compressor power input, the magnitude of which is far less than the latent heat involved in vapor production. Because of this small magnitude, the waste energy of this type of plant can be eliminated as heat energy in the brine and distillate streams, by discharging these streams at temperatures a few degrees above the temperature of the incoming seawater. The economy of this process is favorable, due to the regenerative use of heat, noted above. Practically, it has been found that the electrical energy required is about 50 kWh(e) per kgal, which represents an energy cost of $0.50 per kgal when the power cost is $0.01 per kWh(e). Variations and practical details of vapor compression plants are discussed in Chapter 8.

5. Solar Distillation

The foregoing distillation processes make use of heat energy from the combustion of a fuel. In the case of the two thermal distillation methods the heat energy was transferred to steam and this steam

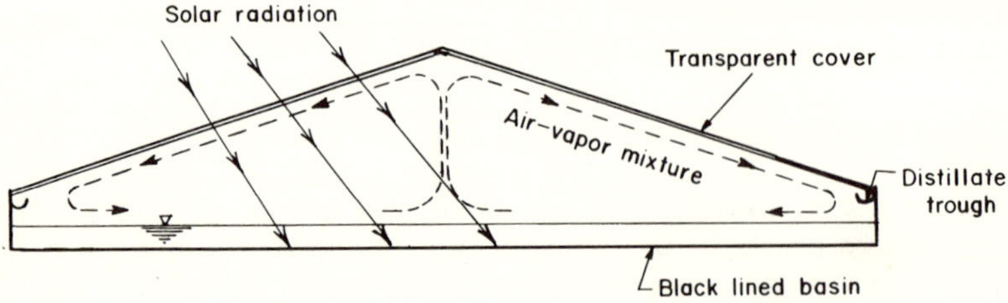

FIG. 4-6. Schematic cross section of a solar distillation still.

condensed in the distillers. In the case of the vapor compression method, the heat energy was converted to mechanical energy before being delivered to the distiller. The fuel costs in all of these cases represent a substantial fraction of the cost of producing distilled water. Because of this there has been considerable effort directed toward the use of solar energy—a "free" form of energy except for the cost of collecting it.

The device most commonly used for solar distillation is the simple solar still, shown schematically in Fig. 4-6. It consists of a black-lined shallow basin of saline water enclosed by a transparent cover with sloping sides. Solar radiation passing through the transparent cover is absorbed by the water and the black basin liner. This radiation is changed to heat by absorption, which serves to warm the water. The warm water partially evaporates and humidifies the air above the surface, thereby reducing the density of the air and causing it to circulate upward. The moving air comes into contact with the inner surface of the relatively cool transparent cover, and part of the humidity condenses thereon. The liquid condensate forms a film and flows to the base of the cover, from where it drips into the condensate trough and is conducted to the outside of the enclosure. The cooled air returns to the surface of the warm water to repeat the process of humidification. The circulation of air is thus due entirely to free convection.

Solar distillation is particularly applicable to isolated desert and island areas where saline water may present problems and small production rates are required. The cost of distilled water using this scheme is about \$3.00 per kgal in spite of the fact that the energy is "free." The quantity produced averages about 1 gallon for each 12 ft^2 of basin area. Specific details of design and cost may be found in Chapter 10.

B. Freeze Separation

Freeze separation, sometimes referred to as cold distillation, is a process which makes use of the fact that ice crystals, formed from saline water, do not contain any of the dissolved salts. If such crystals can be separated from the residual brine and then melted, the resulting water will be quite pure. The principal problem in the development of this process has been the separation of the ice from the brine.

The freeze-separation process involves three discrete steps in obtaining freshwater from a saline source:

1. Ice formation by means of the extraction of heat from the saline solution
2. Removal of the ice so formed from contact with the saline liquid
3. Melting of the ice by the addition of heat to obtain liquid water

The extraction of heat to produce ice at freezing temperatures requires some form of refrigeration, which uses as refrigerant either the water itself or some fluid, such as isobutane, which has both a very low solubility and a low boiling point. The one form of freeze separation which has reached a commercial status is a direct freezing process, and is shown schematically in Fig. 4-7. In this scheme, a vapor compressor is used to remove the vapor from the freeze chamber, thereby extracting the heat from the water and causing part of it to freeze. From the diagram it will be seen that the seawater is precooled before being sprayed into the freeze chamber. The freeze chamber is maintained at a very high vacuum (vacuum pump not shown), so that vapor will form at the freezing temperature (25°to 28°F for seawater). Formation of the vapor causes ice to form in the proportion of nearly 7 lb of ice for 1 lb of vapor. The water itself serves as a refrigerant. The ice is in the form of small platelike crystals and these together with brine are forced into the wash column by the slurry pump. The ice crystals rise in the wash column due to buoyancy and the brine is pumped out of the bottom of the column. The crystals, reaching the top of the wash column, are pushed down a chute into the melter. A small amount of the product water (approximately 5%) is sprayed over the surface of the ice in the top of the

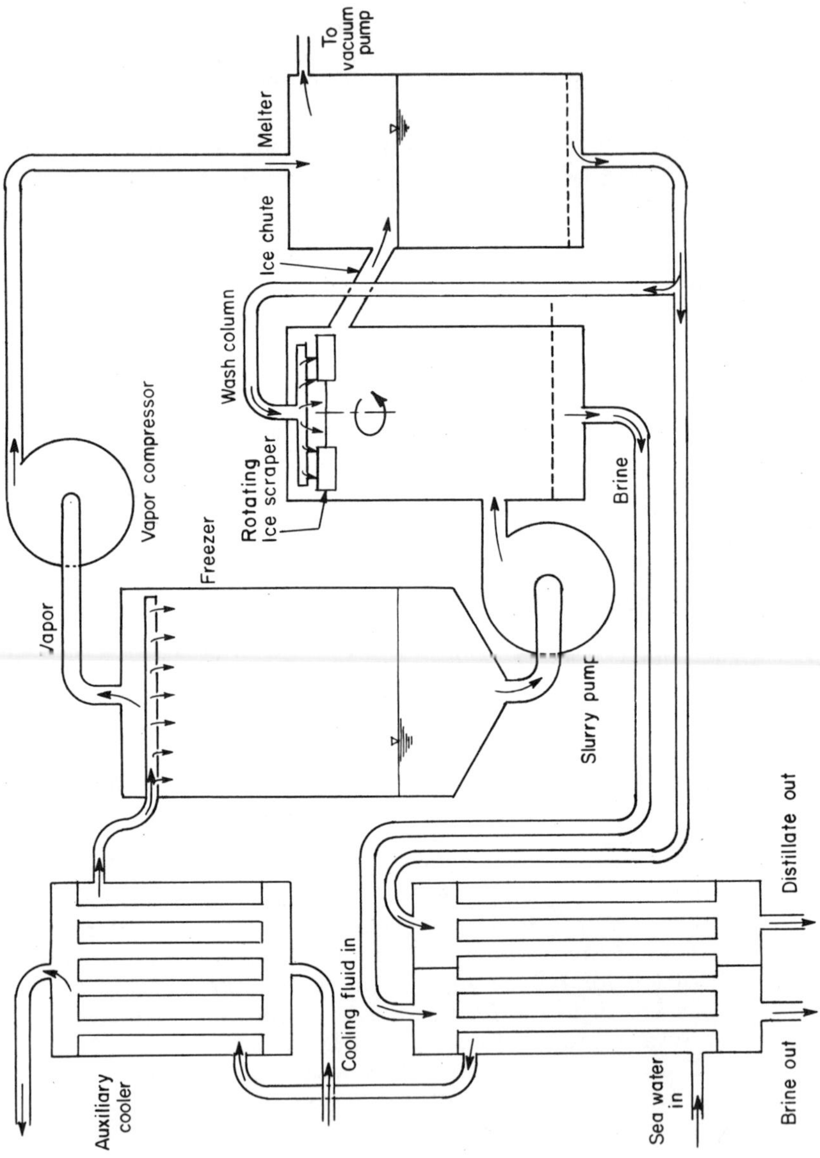

FIG. 4-7. Schematic cross section of a freeze-separation plant.

wash column before the ice is pushed down the chute, so as to rinse away the brine which tends to cling to the crystals. This wash water trickles down through the ice column and is discharged with the brine. Compressed vapor from the freeze chamber is discharged into the melter, where it surrenders its latent heat to cause melting of the ice.

The process described may be regarded as a vapor compression distillation process in parallel with a freeze-separation process. The distilled water produced by the combination of melted ice and condensed water will be less pure than that of the distillate from the vapor compression process because of the difficulty of freeing the ice crystals from the brine. At first glance it would appear that very little energy is required to operate the process, since only one-seventh of the total product water passes through the vapor compressor, the balance of the water coming from the melted ice. However, to the compressor power must be added the power required to produce the vacuum, to circulate the water and ice through the system, and to pump the waste heat of the process up to the ambient temperature of the cooling water. All these total power requirements result in approximately the same amount of energy as that required for a vapor compression installation.

The waste heat mentioned in the previous paragraph includes the energy inputs to the several pumps and motors needed to operate the plant and the inleakage of heat from the surroundings, the latter being relatively small. Elimination of this heat from the system is necessary in order for the plant to continue to operate at constant temperature. This waste heat cannot be carried out by the waste brine and distillate because these fluids are lower in temperature than the incoming seawater. For this reason a heat pump is required. Figure 4-7 shows an auxiliary refrigeration unit arranged to cool the incoming seawater. The heat energy absorbed from the water is sufficient to offset the waste heat of the system, and the heat so absorbed is passed on to the cooling water in the condenser of the refrigerating unit and ultimately to the sea or surroundings.

The freeze-separation process described above has the basic difficulty that the vapor pressure of the water in the freezing chamber is only about 0.12 in. of mercury. This low pressure requires very effective vacuum equipment, and the low density of vapor at this low pressure requires an extremely large vapor compressor to deliver the required weight of vapor to the melter. For this reason alternative schemes using refrigerants other than water will be dealt with in Chapter 11, in the search for more efficient application of the freeze-separation concept.

C. Hydrate Separation

The hydrate separation process is quite similar to the freeze-separation process described above. The freeze chamber is replaced by a reactor chamber in which the pressure and temperature are accurately maintained at values such that hydrate crystals will form. Hydrate materials have the advantage of forming crystals (freezing) at temperatures much higher than that of seawater ice. A few of these materials are listed in Table 4-1. The hydrates of these materials are known as clathrates, since they form cagelike structures with one molecule of the hydrating agent serving as the "keystone" of a crystal containing several molecules of water. This latter number varies with different hydrates; e.g., it is 17 for propane, a hydrate frequently used in this process, but only 8 for methyl chloride. A further discussion of hydrate separation will be found in Chapter 11.

III. PROCESSES UTILIZING PROPERTIES OF MEMBRANES

The processes discussed up to this point involve the separation of freshwater from saline water by phase changes of the water compo-

TABLE 4-1

Comparison of Hydrating Agents[a]

Material	Reactor conditions	
	Temperature (°F)	Pressure (Psia)
Propane	33.2	69.7
F-21	39.0	12.2
F-12Bl	41.4	20.8
F-142B	46.8	28.6
Methyl bromide	49.4	18.7
F-31	56.1	35.7

[a]Adapted from F.A. Briggs and A.J. Bardulm, Fluorocarbons in the Hydrate Process, A.C.S. Advances in Chemistry Series, No. 38, 1963, pp. 190-199.

nent, i.e., water in solid or vapor form is separated from the liquid solution. Any energy required for separation will therefore be proportional in amount to the total amount of freshwater produced, and will be relatively independent of the salinity of the feed stream. In contrast, there are several processes in which either the unwanted salts or pure water are removed from a saline solution by the use of chemicals or selective membranes. These latter processes are most applicable to waters of low salinity where the energy required to produce a given quantity of potable water is much less than where that same quantity will be produced through phase change, as in distillation or freeze- or hydrate-separation. Although not as pure as distilled water, its quality can be made relatively free of dissolved salts. In electrodialysis and ion exchange, both of which withdraw ions from the product stream, the energy required for desalting is proportional to the amount of ions removed, and can be relatively low if the initial salinity is 5000 ppm or less. In reverse osmosis the energy required varies with the osmotic pressure of the solution, which increases with salinity.

It is important before describing the individual membrane processes to define the terms osmosis and dialysis. Both of these terms refer to situations where a membrane serves as a barrier separating two aqueous solutions of different concentrations. Osmosis is the transmission of water through a membrane from the less concentrated solution to the more concentrated one. Dialysis is the diffusion through the membrane of dissolved salts, the diffusion transferring ions from the more concentrated solution to the less concentrated one. The degree to which these transmissions and diffusions take place differs from one membrane to another, but on the whole, osmotic membranes transmit relatively few dissolved ions and dialysis membranes allow only a small amount of water transfer.

The two processes that utilize membranes for desalination, electrodialysis and reverse osmosis, have become more practical in recent years due to the continuing development of plastic membranes. Where previously sheet plastics had been neutral to the electrical charges of dissolved ions in the solutions with which they came into contact, electrodialysis membranes can now be made to contain either positively or negatively charged sites which attract ions carrying electrical charges of the opposite polarity. Synthetic osmotic membranes have also had little application until very recently because, although they have long been used for laboratory measurement purposes, they have not had the mechanical strength or the property for transmissivity of water necessary to accomplish desalination by means of reverse osmosis. Development of both types of membranes is continuing and improvements in their properties can be expected, so that in the future they should offer increased effectiveness and economy for these specialized desalting processes.

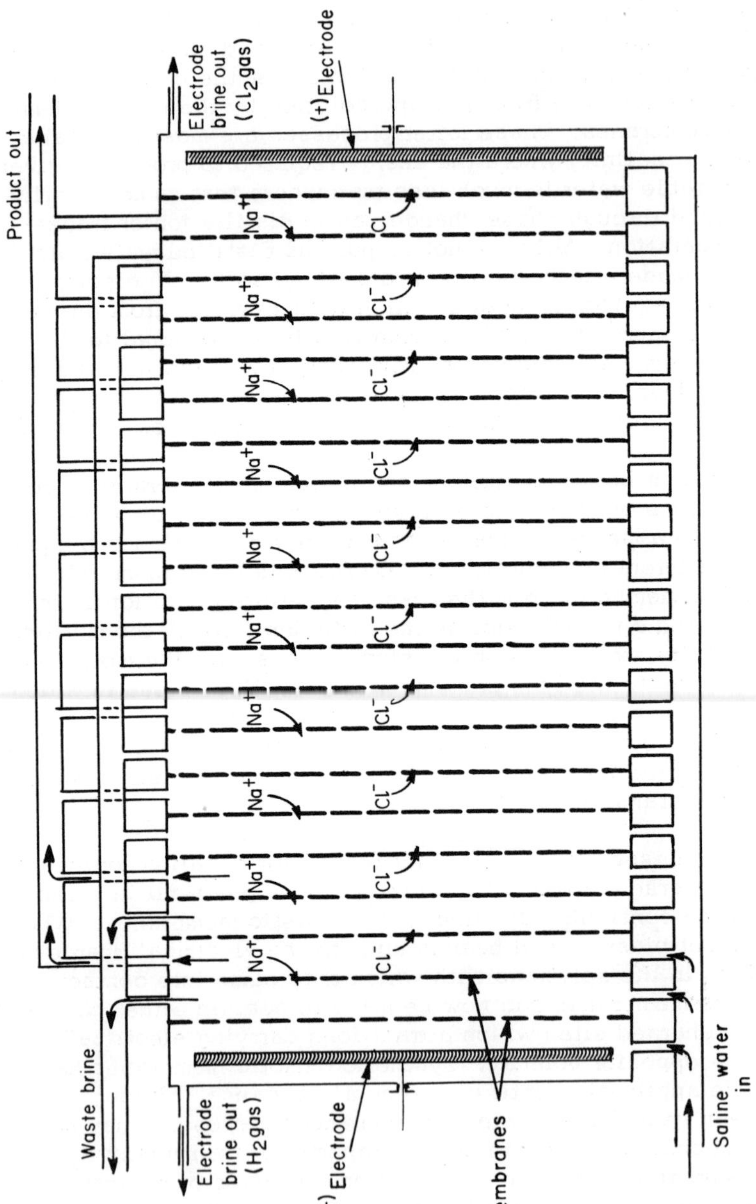

FIG. 4-8. Schematic cross section of an electrodialysis plant.

A. Electrodialysis

In electrodialysis the saline water is divided into two streams and the unwanted ions are withdrawn from one stream, which becomes the product stream, and added to the other waste or concentrate stream. The transfer of ions from one stream to the other is accomplished by using a dc electrical potential transverse to the water flow which causes the ions to migrate across the stream and through the membranes. Selective membrane sheets, perpendicular to the electrical potential and parallel with the direction of flow, act as one-way check valves to prevent the reentry of the ions into the product stream. These membranes are set up in pairs, one membrane permitting positive ions to pass through itself while repelling negative ions, and the other member of the pair allowing negative ions to pass through, repelling positive ions.

An electrodialysis stack consists of many pairs of membranes placed parallel to each other with adjacent membranes held apart by thin spacers (about 0.06 in. thick), so that water can flow over the surfaces of the membranes. Figure 4-8 shows a schematic cross section of an electrodialysis stack containing ten pairs of membranes. The electrical current flows horizontally between the two electrode plates on the outer side of the membrane stack, distributing the dc electrical current across the entire surface of the membranes. The water streams, flowing across these membranes in alternate channels, are manifolded together, separating the water from these two sets of channels into product water and concentrate.

From the above, it can be seen that in the electrodialysis process electricity is carried by the migration of ions. Because of the selective nature of the membranes, the ions are removed from one set of membrane passages and discharged into the other set as indicated by the Na^+ and the Cl^- ions in Fig. 4-8. Thus one of the two streams becomes desalted and the other becomes saline, thereby carrying away most of the ions. It must also be realized that there are separate reactions at the electrodes, where oxidation and reduction are necessary for passage of electrons between electrodes and solution. Depending upon the metals used for the electrodes, there may be corrosion of the metals, but the chief electrode reactions are generally the formation of gases such as hydrogen and chlorine, as indicated in Fig. 4-8.

The electrodialysis process cannot remove all of the ions from the product stream because such removal would require very high voltages to overcome the great electrical resistance of highly pure water. From a practical viewpoint, the freshwater stream leaving an electrodialysis

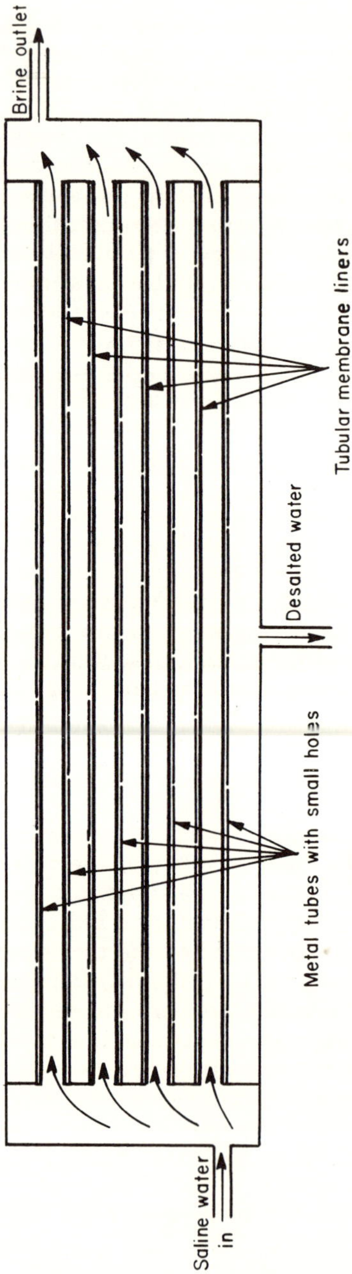

FIG. 4-9. Schematic cross section of a reverse osmosis cell.

plant is so controlled that the final salinity will be approximately 300 to 500 ppm. Experience has shown that passage of saline water through one electrodialysis stack will produce water of about 50% of the salinity of the feed water entering the stack. Thus if the source water has a salinity of 4000 ppm and the product water is to have a salinity of less than 500 ppm, it would be necessary to pass the water through four stacks in series to achieve the desired result. On the other hand, if seawater is the source, with a salinity of 35,000 ppm, the production of water with not more than 500 ppm salinity would require the passage of water through seven stacks in series. The capital cost of an electrodialysis plant is proportional to the number of stacks involved, so that the multiplicity of stacks needed in series for the desalting of highly saline water increases the cost quite rapidly. The electrodialysis method can be cheaper than distillation when the source water has a total salinity of about 4000 or 5000 ppm, but is more costly than distillation when used with seawater. An extended discussion of the background for this conclusion will be found in Chapter 13.

B. Reverse Osmosis

Another membrane process for desalination which has been developed in recent years is known as reverse osmosis. It takes its name from the fact that in this process water is made to pass from the more concentrated solution to a less concentrated one; which is the reverse of the osmosis phenomenon described earlier. The force necessary to accomplish this is the application of pressure greater than the osmotic pressure of the saline solution. While a number of membrane materials seem to have effective characteristics for use in the reverse osmosis process, those which have proved most successful are made of cellulose acetate. Membranes are cast under carefully controlled temperature conditions in sheet or tubular forms, with a thickness of about 0.004 in. The casting solution consists of cellulose acetate, acetone, water, and a small amount of a salt such as magnesium perchlorate. After casting, this latter salt is leached from the membrane by soaking it in water before being assembled in the apparatus. The routines for casting the membranes and leaching the salts have been the subject of much research, inasmuch as the temperatures employed during the manufacturing operations have been found to affect the water transmissivity property as well as the selective properties.

Figure 4-9 shows a schematic cross section of one form of a reverse osmosis cell. The device shown is a shell and tube arrangement with the saline water on the insides of the tubes. The outer metal tubes are perforated at intervals with very small holes, and lined with a porous material. Membrane tubes are placed inside the porous lining and sealed to the outer tubes at the ends. Water flowing through the membrane tubes under very high pressure, about 600 psi for brackish water, filters through the membrane wall. The porous liner conducts this water along the metal tube to the nearest perforation, through which the water passes and is collected in the shell.

While reverse osmosis devices are presently used commercially, the amount of practical experience with them is limited. It appears that for saline waters of 6000 ppm or less, the pressure drop across the membrane will be of the order of 600 psi and will produce a water flux of 10 to 20 gpd per ft^2 of membrane with a salt rejection of 90 to 95%. The energy requirement is about the same as that for electrodialysis. Further details of construction and operation will be found in Chapter 14.

1. Terminology Controversy

Recently certain authors have preferred the term "hyperfiltration" for this process, rather than reverse osmosis. In this book the original terminology, "reverse osmosis," has been retained. The reasoning behind the preference for hyperfiltration lies in the meaning of its prefix, "hyper," which signifies something above, over or beyond the usual norm, namely, in this case the effective filtration caused by the application of a pressure greater than the osmotic pressure of the solution. Since it is the osmotic property of the membrane that is involved in the process, and which is unspecified in the term "hyperfiltration," it seems more descriptive to refer to the process as "reverse osmosis."

A second reason for preferring the latter designation arises from the fact that filtration processes normally result in the collection of suspended material into a solid cake, while reverse osmosis merely concentrates the dissolved solids in solution without completely dewatering them. The reason for this difference is that ordinary filtration of a given volume of liquid suspension can be accomplished with very little increase of pressure on the liquid. In contrast, since osmotic pressure increases with concentration, the rate of flow through an osmotic membrane from a closed volume of solution under constant pressure will decrease as the concentration, resulting from the removal of the product water, increases. When the increased concentration raises the osmotic pressure to a value equal to that of the applied constant pressure, the flow through the osmotic membrane will stop. Thus, for any given applied pressure there is a certain concentration of the brine at which the flow of water through the membrane will cease.

For the range of pressures applicable to reverse osmosis, the limiting concentrations are very far from the complete dewatering associated with filtration. For these two reasons, therefore, the term "reverse osmosis" seems more appropriate, and is used throughout this book when referring to this method of desalination.

IV. PROCESSES UTILIZING ION SELECTIVE PROPERTIES OF SOLIDS AND LIQUIDS

A. Ion Exchange

Ion exchange has been used extensively for softening water and to a lesser extent for "polishing" high purity water. The softening process originally used a natural sandlike material, called zeolite. This was a double salt, such as hydrated sodium aluminum silicate. When "hard water" containing magnesium and calcium ions was passed through a bed of zeolite, the sodium ions would change places with the magnesium and calcium ions so that the solution leaving the bed would contain only sodium ions. The unwanted magnesium and calcium ions precipitated upon the zeolite crystals in the bed. When the zeolite had lost all of its sodium ions and was therefore saturated with the unwanted ions, it could be regenerated by passing concentrated sodium chloride solution through the bed, thereby reversing the exchange process and restoring the zeolite to its original sodium form. It will be noted that this "softening" process does not reduce the total ionic concentration, but does eliminate the ions which cause hardness.

A different application of ion exchange is referred to as "polishing," and has been particularly applied to produce high purity water for feedwater for high-pressure steam generators. The feasibility of this means of desalination is being continually improved through the development of new ion-exchange materials and resins. One reason for the effectiveness of these granular exchange materials is that they can be regenerated by acids and bases, so that the exchangeable ions left on the grains after regeneration are hydrogen and hydroxide ions. When water containing unwanted ions is passed through beds of such exchangeable materials, the unwanted ions are exchanged for hydrogen and hydroxide ions. The latter immediately unite to form water molecules and leave the plant in a "polished" water stream, stripped of all the dissolved ions.

Figure 4-10 shows the ion-exchange polishing process schematically. Two columns are indicated, one containing a cation-exchange

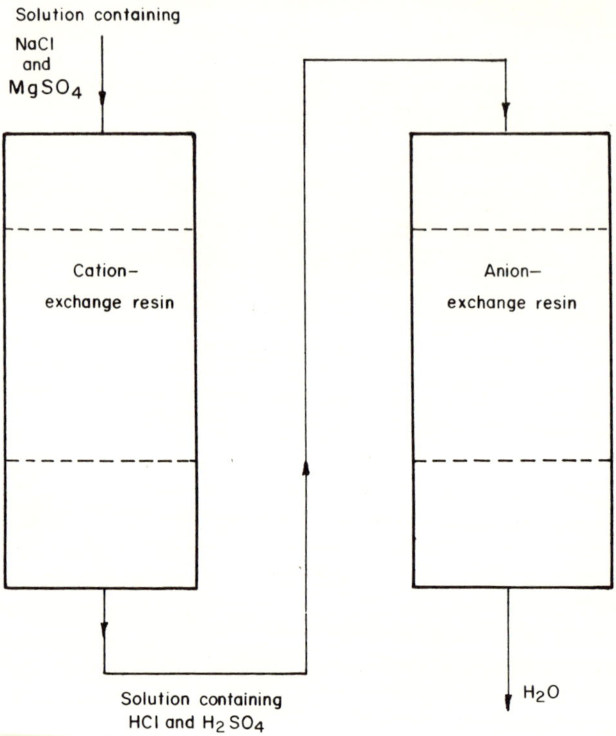

FIG. 4-10. Schematic cross section of the ion-exchange polishing process.

resin with hydrogen ion as the changeable one and the other containing an anion-exchange resin with hydroxide exchangeable. Water containing NaCl and $MgSO_4$ flows downward through the cation resin. The sodium and magnesium ions exchange with the hydrogen ions in the bed, and the water leaves the bed without these cations but with an increased hydrogen ion component. This low pH water passes into the anion-exchanger column where the sulfate and chloride ions exchange with the hydroxide ions. These hydroxide ions combine with the excess hydrogen ions from the first column to form water molecules, so that the water leaving the second column has been stripped of all dissolved materials as compared with the incoming stream.

While this method produces high purity water and could be used to desalt seawater, the cost of regenerants for large quantities is prohibitively high. For example, seawater containing 0.604 g-equiv per liter of positive and negative ions would require an equal amount of hydrogen and hydroxide ions. If obtained from sulfuric acid at \$30 per ton, the cost of the hydrogen ion alone would be \$3.70 per kgal of desalted water. This illustrates the exorbitant costs for chemicals required for the direct use of ion exchange for desalination. Further information on variations of this process will be found in Chapter 12 where it will be shown that this process is only justified when relatively small quantities of high purity water are required.

B. Solvent Extraction

Where ion exchange uses the selective properties of certain solids to accomplish desalination, solvent extraction makes use of the selective properties of certain liquids. There are some liquids which absorb water much more readily than salts and have suitable temperature-solubility relations. Typically, these liquids dissolve water in the amount of 30 to 40% by weight at ambient temperature, and should retain something less than 10% at temperatures of 150°F or more. As a result of the testing of many materials, certain secondary and tertiary amines were found to be the most suitable. It was found possible to use extraction temperatures close to ambient with mixtures of pairs of amines.

The process used for desalination includes three phases: (1) the extraction phase in which the saline water passes through an extraction column countercurrently to the solvent liquid; (2) the separation phase, in which the saturated solvent is heated and the two liquids separated by decantation; and (3) the stripping phase, in which residual solvent is recovered from both brine and product streams. Difficulties with the stripping phase have been so great that research on this scheme was discontinued some years ago.

V. SUMMARY

In the field of desalination a multitude of separatory processes are possible, all of which depend upon the physical and chemical properties of water, but only a small number of these have been seriously considered for desalting large amounts of saline water. Those processes which have been developed to a commercial status, or have been the subject of serious research or pilot plant investigation are described in this chapter. They may be classified in a number of ways, but the classification preferred in this book is based on the phenomena which causes the separation, i.e., phase changes, selective properties of membrane surfaces, and selective absorption capacities of solids and liquids.

The phase change of water from liquid to vapor is the basis of all forms of distillation, the oldest large-scale system being multiple-effect distillation. For very large plants this type has been supplanted by multiple-stage flash distillation. These two methods differ in the manner in which the heat of condensing steam from an external source is reused to produce additional pounds of distillate. A third method of distillation, vapor compression, uses mechanical rather than heat energy to accomplish the regenerative use of heat in its production of vapor. In contrast to the above, freeze separation and hydrate separation utilize the phase change of water from liquid to solid. These two latter processes require the separation of the ice or hydrate from the brine in which it is formed, and the problem of such separation has been difficult to solve. As a result neither method is available on a truly commercial basis.

Processes which depend upon selective membrane properties do not involve a phase change of water, and are, therefore, attractive from the viewpoint of the energy required for separation. Electrodialysis makes use of membranes which are selective with respect to ions which can pass through them when a dc electrical potential is imposed across the membrane. With positive and negative membranes alternating in stacks between electrodes, water passing between the membranes is split into two streams, one of nearly pure product water and the other the concentrate which carries away the ions removed from the product stream. Reverse osmosis uses membranes which are relatively impervious to ions and allow water to be squeezed through them by high water pressures. Both of these membrane processes are used primarily for desalting water whose initial salinity is not more than about one seventh the salinity of seawater.

Other processes which do not require a phase change of the water are ion exchange and solvent extraction. Ion exchange depends upon the exchange capacities of certain special resin beads, and solvent extraction depends on the preferential absorptivity of a liquid for water molecules rather than salt ions. In both cases the saline water is passed through one or more columns of active material where the intended process occurs. Ion-exchange columns discharge the relatively pure product water, while solvent extraction columns discharge brine. The ion-exchange columns must be regenerated to remove the undesirable ions. Solvent extraction columns must be heated or cooled to thereby decrease the solubility of water and permit the product water to be separated from the solvent.

Chapter 5

PROBLEMS COMMON TO DISTILLATION

The several processes for distilling water have been described in qualitative terms in the previous chapter. Each of these processes will be fully discussed in one of the chapters which follows. This chapter will attempt to deal specifically with some basic matters common to all of the distillation processes, particularly heat energy relationships, pretreatment of feedwater, entrainment separators, and post-treatment of distilled water.

I. HEAT-ENERGY RELATIONSHIPS

A. Heat Rejection Requirements

1. Basic Requirements for Distillers Operating with Heat and Mechanical Energy

Basic for determining an effective and economical desalination plant design is an understanding of heat-energy relationships, i.e., the balance between heat-energy input and reject. Axiomatic, the heat-energy input must be equaled by heat rejection in order for a steady-state condition to exist within a plant. The problem of heat dissipation has been introduced in Chapter 3, but will be elaborated

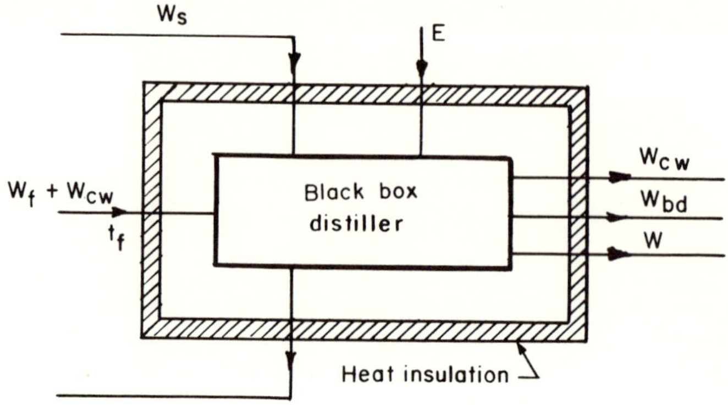

FIG. 5-1. Idealized distillation plant. Note: $Q_s = W_s h_{fg}$.

upon here in a more quantitative manner. The chief aspect with which heat rejection is concerned in this section involves the determination of the amount of cooling water which must be provided in desalination plants in order to maintain the steady input-reject balance. Figure 5-1 is a schematic diagram illustrating the "idealized" process effecting the heat-energy input-reject balance in any of the distillation processes, with the possible exception of solar distillation. In this figure the basic distillation process is represented as housed in an "insulated black box" enclosure, into which saline water flows at a steady rate and from which distilled water, cooling water, and brine blowdown leave at steady rates, such that there will be no storage of either water or salts within the enclosure; i. e., the rate of the former is equal to the sum of the rates of the latter, as is indicated in Eq. (1):

$$W_f + W_{cw} = W_{bd} + W_{cw} + W \tag{1}$$

where W is the weight of liquid in a given time, and subscript f is the feed to separatory system, cw is cooling water, and bd is brine blowdown (no subscript indicates product).

It also follows that if temperatures of these four streams are to remain constant with time, energy that is added to the plant will have to be accounted for by temperature elevations in the three exit streams above that of the incoming water. Using the above flow-rate equation, and assuming zero accumulation of energy within the system, the total energy balance can be written as follows:

$$Q_s + E = Wh + W_{bd}h_{bd} + W_{cw}h_{cw} + W \cdot \overline{Wk_0} - (W_f + W_{cw})h_f \tag{2}$$

where Q_s is the heat input in Btu per unit time, E is mechanical and electric input Btu per unit time, h is enthalpy, and $\overline{Wk_0}$ is energy of separation.

For the purpose of this discussion it can be assumed that the specific heats of the water in all streams are equal to 1.00 and that changes in enthalpy are equal to the product of specific heat and temperature change. Equation (2) can then be written as:

$$\frac{(Q_s + E)}{W} = c(t - t_f) + \overline{Wk_0} + \frac{c(t_{cw} - t_f)W_{cw}}{W} + \frac{c(t_{bd} - t_f)W_{bd}}{W} \quad (3)$$

where c is the specific heat at constant pressure and t is temperature.

The quantity Q_s can be expressed in terms of the weight of steam used, its equivalent latent heat, and the GOR, based on steam alone. As an example of the application of this equation, the conditions at the distillation plant in Freeport, Texas may be used. Analysis of performance data given in the second annual report [1] shows:

W_{bd} = brine blowdown rate = 0.333 lb per lb distillate

t_f = seawater inlet temperature = 78°F

t = distillate temperature = 93°F for 51.2% of product and 109°F for 48.8% of product (note that condensate is mixed with distillate and cooled to 109°F)

t_{cw} = cooling water temperature = 86°F

t_{bd} = brine blowdown temperature = 117°F

GOR = 11.62 lb distillate per lb steam

Δh_s = heat available from steam = 1195.2 Btu/lb

$\overline{Wk_0}$ = minimum work of separation = 5.5 kWh/kgal

E = electrical energy input = 11.25 kWh/kgal

c = specific heat = 1.0 Btu/(°F × lb)

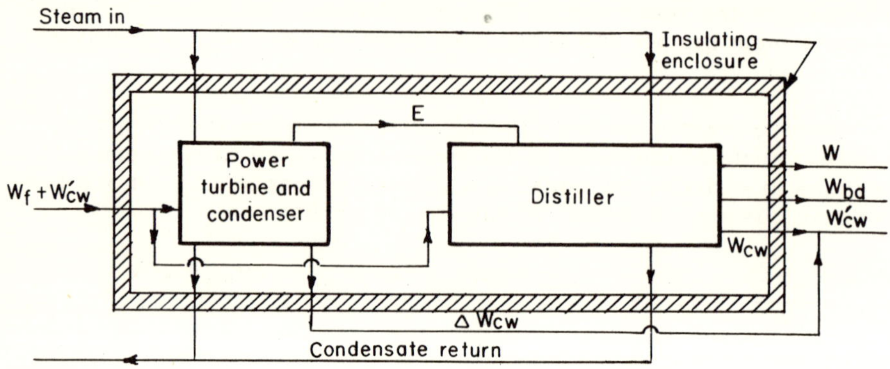

FIG. 5-2. Idealized distillation plant combined with electric generator.

Inserting these values in Eq. (3) yields the ratio of $W_{cw}/W = 8.35$. This figure is to be compared with the reported cooling water rate of 8.02 lb per pound of product distillate. The difference between these two values may be assigned in part to losses in heat from the several heat exchangers, such loss not having been included in Eq. (2). However, since the computed and measured values lie within a 10% differential, the agreement is sufficiently close to demonstrate the method of estimation.

This illustrates the method of calculating the rate of the cooling water needed to maintain steady heat flow within the plant. The foregoing arguments, based on the concept of Fig. 5-1, assumes that the source of the mechanical energy is unimportant from the viewpoint of heat dissipation. The important fact is that the pumps and other devices using mechanical energy are in contact with the water and, consequently, any mechanical energy used by them will ultimately appear as heat energy in the water. The source of the mechanical energy could be as diverse as an external electrical distribution system, or a backpressure turbine placed in the steam line between the boiler and the brine heater of the distiller, and still make no difference in the heat dissipation requirements of the plant.

2. Additional Cooling Water Requirements Due to Power-Generating Units

Besides the above conditions under which mechanical energy is delivered, another condition obtains in some plants, and results in an added load on the cooling system. This condition is exemplified by the schematic diagram of Fig. 5-2, which represents a plant in which one steam generator is used to supply separately a condensing-type turbine and a distiller. The condensing turbine generates the mechanical power for the plant but, unlike the noncondensing or backpressure turbine mentioned in the preceding paragraph, discharges its steam to a condenser and not to the distiller. The condenser requires cooling water, so that the amount of cooling water must be increased above that needed for the distiller plant alone. The amount of additional cooling water involved can be estimated on the basis of several assumptions. The amount of cooling water for a condensing turbine depends on the steam conditions, the temperature of the incoming water supply, and other factors, but may be approximated as being about 25 times the weight of the steam to be condensed. The amount of steam required to produce the needed mechanical energy for a multiple-stage flash-type unit is roughly one-sixth that needed for the brine heating section of the distiller. Thus, the additional cooling water required by the condensing turbine will be about 25/(6 GOR), with the GOR based on the steam required by the brine-heating section of the distiller plant. In the particular case of a plant with a GOR of 12, the added cooling water would amount to a little more than one-third of the distillate production rate. The pumping machinery would need to be increased in capacity to deliver the additional cooling water, and the disposal scheme would need to be enlarged above the distiller-only size to accommodate the additional warm water. In Fig. 5-2 the cooling water entering the plant is shown as $W_f + W'_{cw}$, and is greater than the amount of water needed by the distiller by the amount ΔW_{cw}. The condensing turbine and its condenser are shown as included inside the insulating boundary in order to emphasize the need for including the extra cooling load imposed by the turbine generator.

B. Modification of Economy Parameters

In the section on multiple-effect distillation in Chapter 4 the evaluation of a plant economy was shown to be expressed in terms of its "gained output ratio" (GOR) or its "performance ratio" (PR). The GOR was defined as the amount of distillate produced divided by the amount of external steam supplied, i.e., GOR = W/W_s. The PR is the amount of distillate produced divided by the heat input in thousands of Btu's. These ratios are basic indicators of the economy of a plant and, for accurate comparison of two or more plants, must be adjusted to account for differences in the manner in which the needed mechanical energy is supplied.

Mechanical and electrical power for pumping and other services is supplied to different plants in a variety of ways, including the use of electrical power from central stations, electrical and mechanical power from back pressure turbines which serve as pressure-reducing devices for part of the steam used in the brine-heating section of the distiller, and electrical power from condensing turbines using steam from the same source as the distiller. The adjustment of steam rates to account for these various methods of supplying power has not been standardized. The following suggestions are intended to display some of the possible schemes which could be used.

The simplest correction to the GOR or PR is that involved in the use of back-pressure turbines for power supply. Since the steam used by these turbines is discharged to the brine-heating section of the distiller, the effect on the plant economy is that of requiring extra heating of the steam above that required by the brine heater. The amount of this extra heating is the enthalpy change in the turbine. This could be replaced by the weight of the steam used by the turbine multiplied by the ratio of the enthalpy drop through the turbine to the enthalpy of the steam entering the brine-heating section of the distiller.

A second situation requiring an adjustment to the GOR and PR is that in which electrical power is drawn from an external supply, such as that from a central station. The equivalent steam rate for the central station turbogenerators will depend upon the efficiency with which those turbines operate. While in some cases detailed information may be available, in most cases it would be sufficiently accurate to assume that the overall efficiency with which such power is delivered amounts to 25%, so that the mechanical energy used, expressed in heat units and multiplied by four, would give the amount of heat equivalent. This latter figure divided by the latent heat of the steam used in the brine heater section of the plant would give a good approximation of the added steam requirement due to externally supplied power.

The calculations for a third case in which power is supplied from condensing turbines, the steam source of which is the same as that for the brine heater, is quite straightforward. The amount of steam used by such turbines, multiplied by the ratio of the enthalpy of steam delivered to the brine-heating section of the plant to the enthalpy of the steam delivered to the turbine, would be a close approximation to the additional amount of steam that must be included in computing the economy of the distiller plant. These above suggestions are intended only as guidelines for the evaluation of the total steam consumption, or its equivalent, in distillers, but it also serves to emphasize how important it is to determine the manner in which mechanical and electrical power are supplied to plants when economy comparisons are being made.

C. Economy of Dual-Purpose Plants

The foregoing energy discussions have dealt with problems relating to water-only distillation plants, in the design of which is also included supplementary electrically driven auxiliary machinery. In these installations the electric power generated was nominal, and used solely for the operation of the distillation plant itself. Another facet of the energy relationship is introduced by the dual-purpose concept, in which both electric power and distilled water are commercially produced for large-scale distribution from a single source of heat energy.

Technically, a dual-purpose plant is one in which steam from a boiler passes through a power-generating backpressure turbine, and is discharged at suitable pressure and temperature to the heating section of the distillation plant in which the steam is condensed and returned to the boiler feed system. Such an operation is carried on "in series," and must be distinguished from the "parallel" case in which a single steam generator supplies steam to two separate plants—a distiller and a power plant, each of which condenses its share of the steam and returns the condensate separately to the boiler feed system. This parallel arrangement is not regarded as true dual-purpose arrangement in the technical sense noted above, even though a single boiler supplies steam for both power production and water distillation.

A true dual-purpose arrangement is possible because of the different temperatures of steam with which power plants and distillation plants operate. High efficiency power plants require high temperature steam at the level of 1000°F and upward, whereas distillation plants are limited to maximum brine temperatures of the order of 250°F. While steam turbines for electrical power generation are ordinarily designed to expand steam to about 100°F, they can be designed to exhaust at pressures such that the discharged steam will have a saturation temperature of about 250°F, thereby making it suitable for heat supply to the distillation plant. The major disadvantage of such an adjustment is that it reduces the pressure range through which the steam may expand, and thus requires a greater flow rate of steam for a given power output.

Over the years there have been several attempts to apply the dual-purpose in various installations. An early and unique example of such a plant was described in 1929 [2]. Located in Stearns, Kentucky this plant produced power to operate the mining machinery of that community, but also produced 45,000 gpd of fresh distilled water for drinking. The water for the steam boiler was taken directly from the river and given a chemical pretreatment before being sent to the boiler. The Stearns installation was unique in that none of the condensate from the steam used by the reciprocating engines for producing power was reused in the boiler for purified feedwater, but all of the condensate was diverted directly to the community water supply. The preference for using all of the condensate for drinking water rather than for boiler feed was due largely to the poor quality of the river water for potable use. Pretreatment of the water for boiler feed was the more practical because the iron and other minerals in the river water rendered it so distasteful that mixing it in any large amount with the distillate was intolerable. The chief difference of the Stearns plant from later designs is that there was no supplementary distillation plant as such; evaporation was accomplished in the steam boiler and liquefaction in the power plant condenser.

While the above example displays the heat-sharing concept of dual-purpose operation, its simple arrangement is quite inadequate for large-scale designs proposed for future construction. The first plant which demonstrated the modern dual-purpose concept on a moderately large scale was the plant installed at Aruba in 1958 [3]. The source of the heat was an oil-fired steam boiler which produced high-pressure steam for driving two steam turbines producing 12, 500 kW. Exhaust steam at approximately atmospheric pressure was discharged from the turbines to the first effect of a six-effect distiller plant, rated at 2, 500, 000 gpd. In this dual-purpose scheme the plant's utilization of heat energy for power and distillation was arranged "in series," taking advantage of as much economy as possible. However, because of unequal demands for power and water, and of uncertainties as to the reliability of the two-component plant, the design provided alternate steam connections so that the two plants could operate independently and "in parallel." This was accomplished by providing separate condensers for the turbines, and pressure-reducing valves and desuperheating units for reducing the pressure and temperature of the steam enroute from the boiler to the evaporator section of the distiller plant.

Few dual-purpose plants adhere to the true dual-purpose concept in which the operation is completely carried on "in series." Such a plant would require the production of steam at high pressure and temperature, the expansion of this steam in a turbine for the generation of electrical power, and the discharge of the expanded steam to the distillation plant. As intimated earlier, because of unequal demands for power and water, and unreliability of operation due to the complicated interdependence of the dual plants, this simple "in series" design has serious limitations. This has led to modifications in which the operation is a combination of "series" and "parallel" design. The dual-purpose portion of the plant effects a saving on the amount of energy required and thus a proportional saving in cost. The combination, in addition, effects a saving on the per unit cost of energy, in that a single steam generator serving both a distillation plant and an electrical generating plant would necessarily need to be larger than each of two separate generators serving two plants. The larger the size of the steam generator, the lower generally is the unit cost of steam. This is particularly true for nuclear reactors where a significant increase in size decreases the per unit cost of heat generation at a considerably greater rate than for fossil-fueled steam generators of comparable size. It is because of these drastic cost and energy economies that most proposals for large distillation plants are considered in combination with power-generating facilities.

A widely studied proposal of this dual type of plant is that of the Bolsa Island [4] plant, formerly under consideration by the Metropolitan Water District for construction near Long Beach, California. The arrangement of equipment proposed in this study is shown in Fig. 5-3. This proposed dual-purpose project was to consist of a power generating installation, part of which would operate "in series" with a distillation plant, and the remainder would be a power-only plant. The distillation portion was designed to receive exhaust steam from a backpressure steam turbine rated at 490 megawatts (MW). This steam was to be supplied by nuclear reactors which also furnished steam for two power-generating condensing turbines rated at 655 MW each. Figure 5-3 shows this plant as having two relatively independent sections, one of which contains the dual-purpose plant, and the other the two condensing turbines for power generation. Observing the dual-purpose section of this figure, it is seen that the steam from the nuclear reactor is expanded in the high-pressure section of the backpressure turbine, reheated, and then expanded further in the low-pressure section before being condensed in the brine heaters of the distillation plant. From there the condensed steam is passed through a conventional regenerative feedwater heating system involving six stages of heating, and a "polishing demineralizer" before being returned to the nuclear reactor. The steam furnished to the two condensing turbines of the power-only section, on the other hand, is shown flowing merely through the turbines, condensers, and the regenerative feedwater heating system as it undergoes the normal sequence of state changes for steam in any power-only plant of this type.

The basic decision whether or not to build a dual-purpose plant or two independent plants or a combination of the two types of plants depends upon the relative demands for water and power, and the relative costs of the two types of installations. The relative demand for water and power offers serious limitations to the practical operation

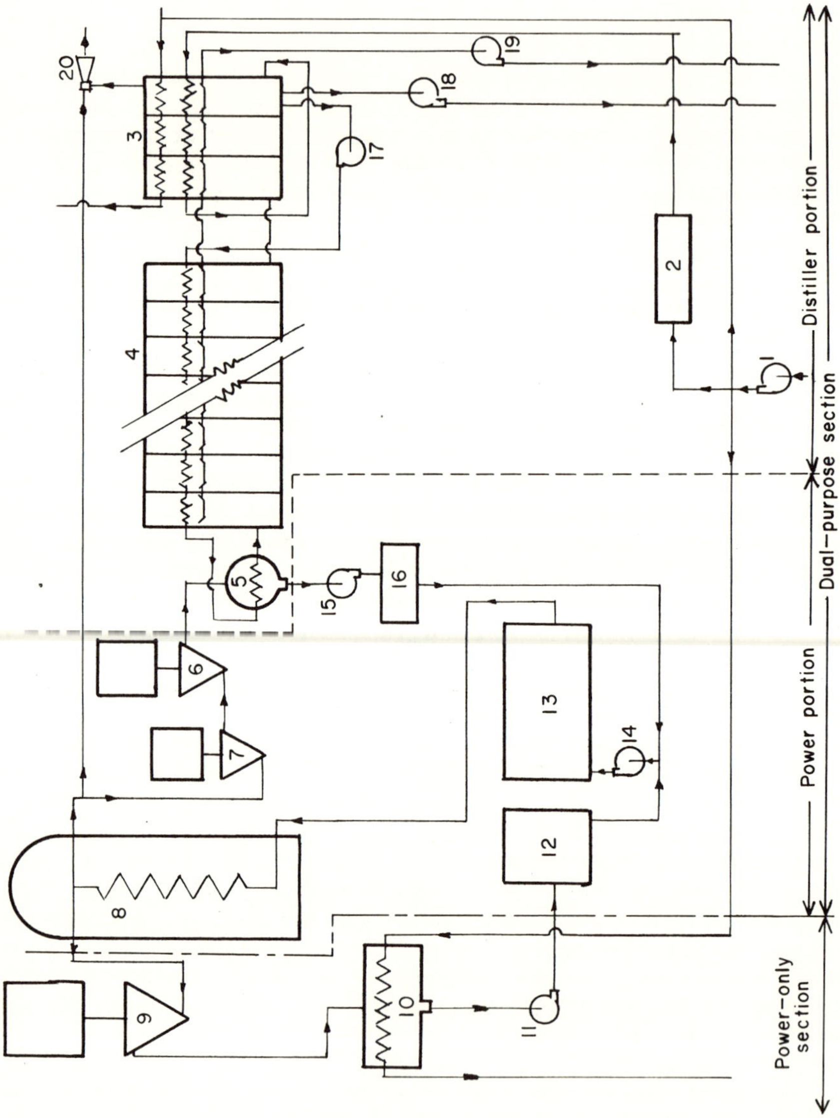
1
2
3
4
5
6
7
8
9
10
11
12
13
14
15
16
17
18
19
20
Distiller portion
Dual-purpose section
Power portion
Power-only section

FIG. 5-3. Combination dual-purpose and power-only plant. (Schematic only.) Arrangement similar to the Bolsa Island proposal. 1. Seawater supply pump. 2. Seawater pretreatment system, includes filtration, chlorination, acidcation, and degasification. 3. Distiller, heat reject section. 4. Distiller, heat recovery section. 5. Brine heater for distiller. 6. Low-pressure section, backpressure turbine. 7. High-pressure section, backpressure turbine. 8. Nuclear reactor for steam production. 9. Condensing-type turbines for power production only (only one shown). 10. Condensers for power-only turbines. 11. Condensate pumps for power-only turbines. 12. Reactor feedwater heating system, low-pressure section, supplied with steam bled from turbines (details not shown). 13. Reactor feedwater heating system, high-pressure section, also supplied with steam bled from turbines. 14. Feed pump for reactor. 15. Condensate pump, for brine heater of distiller plant. 16. Ion-exchange bed for polishing condensate from brine heater. 17. Recirculating pump for distiller. 18. Brine discharge pump for distiller. 19. Distillate pump for distiller. 20. Vacuum pump for distiller.

of a dual-purpose system. In urban networks the 24-hr peak demand for power may be as much as twice the average demand, and even several times the minimum demand. Since there are few facilities for electrical storage as compared with water reservoir storage, the load carried by the power-generating turbines must fluctuate to follow the power demand curve. Because a distillation plant must operate at a nearly constant rate of output in order to produce efficiently, a backpressure turbine supplying steam to it must also operate at constant load. In view of the fluctuation of electrical demand already noted and the lack of facilities for storing electricity, it is obvious that the capacity of the backpressure turbine can be equal at most to the minimum power demand, thus furnishing enough steam for distilling only a fraction of the total water needed by the community.

For instance, in urban areas of the United States the ratio of the average per capita usage of power to water has been found to amount to about 5 kWh/day of electricity to 155 gpd of water. Design studies of relatively efficient dual-purpose plants indicate that production ratios of power to water run about 5 kWh/day to 64 gpd of water, a figure considerably less than the average water demand. Since even the average load produces a ratio less than one-half the daily per capita water demand, the minimum load would produce an even smaller amount. The figures serve to indicate the degree to which a dual-purpose plant can meet the supply-demand needs of a community. If any substantial proportion of the water needs of a community are to be met by such a combination operation, it will be necessary to resort to a water-only plant in addition to the dual-purpose plant. This was not a factor of importance in the case of the MWD proposal for there the main water supply came from other sources. The immediate need was for increased electrical power facilities, and the distiller plant was proposed chiefly for demonstration purposes.

The second basic factor in arriving at a decision between a dual- or two single-purpose plants is dependent upon the relative costs of the two types of installation. The scheme for allocating costs assignable to power and water is not a simple one, but is based on the difference between the amount of energy that will be required in a dual-purpose plant as compared with two plants, one for power and one for water, operating separately. To arrive at this cost difference, annual operating costs are estimated for two electrical power plants which are designed to produce equal amounts of salable electric power. One plant is the dual-purpose plant with backpressure turbines and the other is a power-only plant with condensing-type turbines. The difference in the annual operating costs of these two plants is used as the cost assignable to water production in the dual-purpose plant. A comparison, therefore, of their relative energy requirements is basic for such an estimate.

The MWD study has suggested such estimates based on a dual-purpose plant and a power-only plant, both of which were designed to deliver 1500 MW of salable electricity to the general power network. The dual-purpose plant was to contain a 490 MW backpressure turbine for supplying heat to the distillation plant and two 655 MW condensing turbogenerators for the power section of the plant, totaling 1800 MW. Of this 300 MW excess over the requisite 1500 MW, 100 MW was to be used for the operation of the power plant, 72 MW for the operation of the distillation plant, 28 MW for delivery of the distilled water to the water system some 30 miles away, and 100 MW to supply pumps and other auxiliaries in the general water system supply. The power-only plant of the MWD study was to contain two 800 MW turbogenerators which would furnish 100 MW for operation of the power plant over and above the 1500 MW for the salable power.

It is important in determining cost estimates to compare energy efficiencies and total energy requirements that can be obtained in the dual-purpose and the power-only plants. Such comparisons were made in the MWD report in which the 800-MW condensing turbines were estimated to have a heat rate of 9816 Btu/kWh, a figure about 40% less than the estimated heat rate of the dual-purpose backpressure turbine of 13, 679 Btu/kWh. This unfavorable measure of efficiency for the dual-purpose turbine can be somewhat lessened by comparing the overall weighted average of the three turbines of the dual-purpose plant with the heat rate of the power-only plant. Assuming that the two 655-MW condensing turbines would have approximately the same heat rates as the 800-MW condensing turbines, the weighted average of the three dual-purpose turbines would be 10, 866 Btu/kWh, a value less disparaging than the 13, 679 Btu/kWh when compared with the power-only plant rate of 9816 Btu/kWh.

In contrast to the above unfavorable heat rate efficiency, the dual-purpose plant shows significant total savings in steam flow requirements, with consequent savings in the amount of fuel burned and attendent pollution hazards. These savings in steam can be estimated by comparing the steam flow in the dual-purpose arrangement with that which would be required by the power turbine and distiller separately. The steam-flow estimate for the dual-purpose plant is 7, 699, 000 lb/hr to the back-pressure turbine, and of this, 5, 400, 000 lb/hr is passed on to the distiller. If the backpressure turbine were replaced by a condensing turbine having the same economy as the 800-MW turbines of the power-only plant, and the distiller supplied with steam directly from the boiler, the combined steam consumption of such a combination would greatly exceed that of the equivalent dual-purpose arrangement. For instance, in the above example the 490-MW condensing turbine would require 4, 810, 000 lb/hr and the distiller, 5, 400, 000 lb/hr, a total of 10, 210, 000 lb/hr. When this total is compared with the dual-purpose flow of 7, 699, 000 lb/hr, it can be seen that the dual-purpose arrangement offers a saving of about 25% in this particular case.

On the basis of detailed analyses of the many factors involved, the MWD study report has computed the following cost estimates for both the power-only and dual-purpose plants already described, summarized as follows:

Annual cost of a dual-purpose plant (includes costs of owning and operating)	$46.15 (million)
Annual cost of power-only plant	31.83 (million)
Net cost of water and MWD plant	14.32 (million)
Value of MWD power unrelated to distiller	3.74 (million)
Net cost of water production	10.58 (million)
Unit cost of water (150 mgd for 330 days/year)	0.214/kgal

The resulting low cost of distilled water, noted above, is appreciably less than could be realized in a comparable water-only plant, indicating that the dual-purpose operation has attractive possibilities from the viewpoint of the economies of water production.

II. PRETREATMENT OF FEEDWATER

Pretreatment of feedwater is an important step in the distillation of saline water and is employed to reduce interruptions to plant operation which result from scale formation and corrosion. Scale formation interferes with heat transfer and with water flow, and because this problem is so universal designers frequently make allowance for it by increasing the heat-transfer area sufficiently to offset the increased resistance. Corrosion of the heat-transfer tubing also seems to be inevitable. Allowance for it is made by increasing the thickness of the tube walls above that needed for mechanical strength so that their point of failure due to corrosion will not occur too soon. In addition to these design features, treatment of the feedwater by deaeration, acids, and other additives can also materially reduce the tendency for scale formation and corrosion.

Pretreatment includes such operations as filtration, chlorination, scale control measures, and deaeration. The former two are generally applied to all of the water supplied to the plant in order to prevent the clogging of small passages in the apparatus by solid materials or by organic growths. These operations have become so standardized that they need not be discussed further here, except to note that the chlorination of seawater feed for distillation plants involves the addition of from 0.04 to 5 ppm of chlorine, which results in a residual chlorine content of about 0.5 ppm. This level is considered sufficient to control the organic materials found in most waters. While chlorination and filtration are applied to all the water entering the plant, scale control measures and deaeration to minimize corrosion are applied to only that portion which is fed to the distiller for partial evaporation.

A. Scale Control Measures

Composition of seawater and saline water from other sources have been listed in the tables in Chapter 2. There, it was pointed out that most of these waters include the ionic composition of one or more salts having inverted solubilities in which their saturation solubilities decrease as temperature is increased. The scale-forming salts of this type most frequently encountered are $CaCO_3$, $Mg(OH)_2$, and $CaSO_4$, the latter substance appearing in three different states of hydration. The solubilities of $CaCO_3$ and $Mg(OH)_2$ vary widely with the alkalinity of the solution as measured by the pH value, but the solubilities of the three forms of $CaSO_4$ are nearly independent of the pH value. Thus the schemes used for controlling the formation of scale will be treated in two parts, the first one dealing with the alkaline components and the other with control of the sulfate scale components.

1. Control of Alkaline Scales: $CaCO_3$ and $Mg(OH)_2$

As indicated in Chapter 2, the bicarbonate ions in seawater begin to disintegrate when the water temperature is increased to about 160°F. Continued heating at or above this temperature is likely to result in the precipitation of scales containing $CaCO_3$ and $Mg(OH)_2$, using the carbonate and hydroxide ions produced by the breakup of the bicarbonate ions. Since it is desirable to operate most distillation plants at temperatures above 160°F, various schemes have been developed for preventing the adherence of these scaling compounds to the heat-transfer surfaces, all of which involve the addition of chemical materials as suggested in Table 5-1. The oldest of these schemes is that of adding to the feedwater chemicals which will prevent the scale-forming materials from adhering to the heat-transfer surface, even though they may be precipitated from the solution.

TABLE 5-1

Methods for Controlling Alkaline Scale

Additive	Reference	Amount (ppm)	Max. temp. (°F)
For preventing scale adherence			
Polyphosphate compounds	[5]	4 to 6	190 to 200
Vegetable extract	[6]	10 to 20	350
Nucleating seeds	[5]		
For completely disintegrating bicarbonate ions			
Sulfuric acid	[10]	100 to 200	250
Citric acid	[16]		
Ferric chloride	[5]		
For preventing partial disintegration of bicarbonate ions			
CO_2 gas	[5]		

This results in the formation of a sludge which either passes out with the brine or is discharged as blowdown of some of the water in the heat exchanger. The majority of materials used for this purpose contain sodium polyphosphate as the active ingredient. The exact manner in which the polyphosphates inhibit scale deposition is not completely understood. It is generally referred to as a threshold treatment, since the amount of additive is too small for a stoichiometric combination with any of the unwanted ions. Amounts of additives less than the threshold value result in some scale deposition, the amount of such scale decreasing as the amount of additive is increased and becoming zero at the threshold value. Elliot [5] explored the known facts about the behavior of polyphosphates. He noted that these materials can be chain, branched, or ring polymers and that the spacing of the atoms in these structures is very close to the spacing of the atoms in the crystals of $CaCO_3$, so that the polyphosphate polymers could interweave with the crystal structure of $CaCO_3$ and possibly interfere with its growth or with its adherence to a solid surface. His general conclusion was that more information about the interaction between this type of additive and the dissolved salts is necessary in order to improve the use of such materials. The primary shortcoming of this treatment is that polyphosphates tend to revert to the ineffective orthophosphate at temperatures above 190°F, thus setting a limit to the maximum temperature of the brine.

As can be inferred from Table 5-1 attempts have been made to find other materials which would inhibit scale formation. The vegetable extract tested by Othmer [6] was described as containing ". . .numerous amino derivatives of hydroxy-organic acids. . ." Although somewhat indefinite, his tests seemed to indicate that this extract could be used at temperatures as high as 350°F. Published literature fails to show recent application of this additive and apparently it is unattractive to designers and operators of distillation plants. The use of "nucleating seeds" on the surface of which the unwanted scale components would be precipitated, has been attempted by several experimenters but so far has been used commercially on a large scale only by the Russians, as reported by Chernozubov et al. [7]. They used seeds of chalk and were able to operate at maximum brine temperatures of 220°F.

The limitation of temperature mentioned in connection with the polyphosphate treatment has led to attempts to develop methods of scale control which will permit the use of higher brine temperatures, the most effective of which appears to be treatment of the water with acid. As described in Chapter 2, pretreatment with acid will disintegrate the bicarbonate ion, which is the source of the alkaline scale, into CO_2 gas and water. The CO_2 gas can be discharged by a degassing process, and thereby the scaling hazard is removed from $CaCO_3$. Sulfuric acid is the cheapest source of hydrogen ion, which accounts for its relatively wide usage in modern distillation plants. Citric acid has been proposed because it is possible to ship it in solid form. Potentially, it is a low-unit-cost source of hydrogen ion, when its use becomes extensive.

The action of hydrogen ion on the bicarbonate ion is well understood and has been described in Chapter 2. The basic equation for the reaction is:

$$H^+ + HCO_3^- \rightarrow H_2O + CO_2$$

The CO_2 gas must be eliminated from the water in order to avoid possible reversion of the reaction. Degassing can be accomplished either by a separate degasser or by the deaerator used to discharge the dissolved oxygen in the feedwater. This operation is further discussed in a following section on deaeration. When a separate degasser is used it is expected that about 80% of the CO_2 will be removed in that step, the remainder being removed in the deaerator. The principal costs associated with acid treatment are the costs of the acid and injection equipment plus the cost of degassing the CO_2 from the water. The major component of this cost is that of the acid itself, the entire cost of which amounts to about \$0.04 per kgal with sulfuric acid at the rate of \$30 per ton. The chief advantage of the acid treatment for removing the hazard of alkaline scale formation is that it permits the brine temperature to rise to nearly 250°F, which is the limit imposed by the possible formation of $CaSO_4$ anhydrite scale. The degassing operation for discharging the CO_2 reduces the accompanying corrosion hazard which would otherwise be a serious factor.

The use of ferric chloride to disintegrate the bicarbonate ion is a third additive for controlling alkaline scale formation. When ferric chloride is added to a solution containing a bicarbonate ion it reacts as follows:

$$Fe^{3+} + 3\,HCO_3^- \rightarrow Fe(OH)_3 + 3\,CO_2$$

The effect of the reaction is similar in a way to that accomplished by the addition of acid, except that the $Fe(OH)_3$, being very insoluble, precipitates but does not form scale. Although it does not form scale, it does present a disposal problem for it must be eliminated from the system. The active ingredient, $FeCl_3$, has been used in several plants in the Middle East and Caribbean areas. Some plants produce the ferric chloride by an electrolytic process installed within the plant, but difficulties of operating the process have proved to be so great that apparently no new installations of the process have been made in recent years. The problem of removing the CO_2 gas is the same as for the acid treatment.

A further scheme of doubtful possibility which has been studied experimentally, but not as yet fully developed, attempts to prevent the disintegration of the bicarbonate ion and hence the accompanying scale formation hazard. The following equation expresses the reaction when the bicarbonate ion is exposed to high temperatures:

$$2\,HCO_3^- \rightleftharpoons CO_3^{2-} + CO_2 + H_2O$$

If CO_2 gas is added to the solution in sufficient quantities, the direction of the reaction will be reversed, thereby preventing the breakdown of the bicarbonate ion and the resultant formation of alkaline scale.

Of all of the schemes for controlling alkaline scale formation, the only two that have received wide application are the use of polyphosphate additives and the use of sulfuric acid. The costs of these two methods are approximately equal. The acid treatment seems to be preferred when acid is readily available at low cost, but when it must be imported the polyphosphate treatment takes preference.

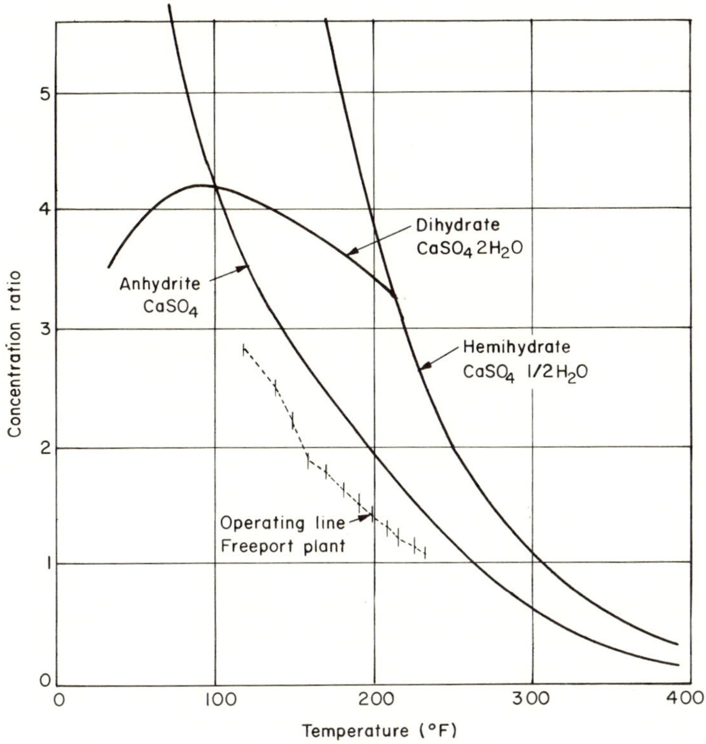

FIG. 5-4. Solubility of scale-forming chemicals in seawater.

2. Control of Sulfate Scales: $CaSO_4$ and Its Hydrates

The temperature at which sulfate scale components begin to precipitate depends on the concentration of the seawater brine and on the particular hydrate involved. The relation between concentration and temperature for the three forms of sulfate crystals found in seawater scales is shown in Fig. 5-4. The coordinates used in these curves are temperature and concentration ratio, i.e., the ratio of concentration of the seawater brine to that of normal seawater. Points above or to the right of the curves represent supersaturation of the liquid and probable precipitation of the scale component. It will be observed that there are separate curves for the anhydrite, the hemihydrate, and the dihydrate. The anhydrite has the lowest temperature range of the three forms and thus is the limiting factor in operating the distiller.

Curves of the type shown in Fig. 5-4 were originally determined empirically by careful experimentation, but recently mathematical methods of calculating these data have been developed. The formulations of Lu and Fabuss [8] and of Marshall and Slusher [9] are particularly significant, since they seem to agree well with experimental data. The curves in Fig. 5-4 were plotted using the data given by Lu and Fabuss [8] for seawater from which the bicarbonate ions have been removed. Similar curves for waters other than seawater are given by McCutchan and Sieder [10], and emphasize the limitation of temperature imposed by the inverted nature of the solubility curves for $CaSO_4$.

Also plotted on Fig. 5-4 is the operating curve of a particular experimental run at the 12-effect distillation plant in Freeport, Texas [11]. This curve represents the concentration-temperature relationship accompanying the increasing brine concentration condition within the successive effects in the plant, brought about by the withdrawal of the distillate. As indicated by this operating line in Fig. 5-4, the design of the plant is so planned that the first effect presents temperature and concentration conditions that lie near but below the high-temperature solubility limit of the $CaSO_4$ anhydrite curve, which is approximately 235°F. As the water proceeds through the several successive lower temperature effects the brine increases in concentration until, in the twelfth effect, it approaches the dihydrate line. Since all of the effects operate below the equilibrium curves of $CaSO_4$, it is not expected that scale of this salt will be formed at temperatures and concentration conditions imposed by the plant design. In this way $CaSO_4$ scale is held under control.

As noted above, the limit on temperature imposed by the anhydrite form of $CaSO_4$ is about 235°F for normal seawater. Since the time required for scale to form is appreciable, some plants have been operated successfully at maximum brine temperatures of 250°F without scale formation when the operating conditions include very fast circulation of the seawater through the high-temperature portion of the plant. However, this practice has not been widely adopted because it is practically impossible to maintain an exactly uniform distribution of water to the many tubes in a heat exchanger. This lack of uniformity leads to scale formation in tubes with low flow rates.

In an effort to most effectively realize the advantage of increased economy of high temperature plants there has been much investigation attempting to reduce the $CaSO_4$ scaling tendency for brine temperatures higher than 235°F. Several proposals, listed in Table 5-2, suggest possible scale preventive measures for the control of scale formation above this critical temperature.

It is noted that the same vegetable extract listed in Table 5-1 is also listed in Table 5-2, since Othmer [6] indicated that it would prevent scale formation up to temperatures of 350°F. Since there is no record of this being used commercially it is sufficient to merely mention it here.

The use of nucleating seeds to collect the scale-forming ions has been the subject of several investigations [7]. As has been discussed earlier, it is conceived that the seeds present nucleation sites of greater attraction than heat-transfer surfaces. The resulting enlarged crystals are filtered out of the brine and returned to the heat exchangers for further augmentation. Because of the inverted solubility nature of calcium sulfate, certain experimenters have attempted to confine the circulation of the nucleating slurries to only the high-temperature heat exchangers, so that the calcium sulfate collected there would not be redissolved by exposure to the lower temperatures of the other section of the plant. It was thought that this method would be attractive because it has been used successfully in plants designed for making salt. However, problems of handling the slurries and distributing the seeds to the essential points in the plant have presented problems of sufficient difficulty that no commercial plants as yet have been able to reach the desired higher temperatures above the saturation curve.

Besides the use of additives, there have been many attempts to prevent the formation of calcium sulfate scales by removing one or

TABLE 5-2

Methods for Controlling Calcium Sulfate Scale

	Reference	Amount (ppm)	Max. temp. (°F)
Additives for preventing scale adherence			
Vegetable extract	[6]	10 to 20	350
Nucleating seeds	[7]		
Schemes for softening saline water			
Ion exchange	[12]		
Lime-magnesium-carbonate	[5]		
Reverse osmosis	[13]		320

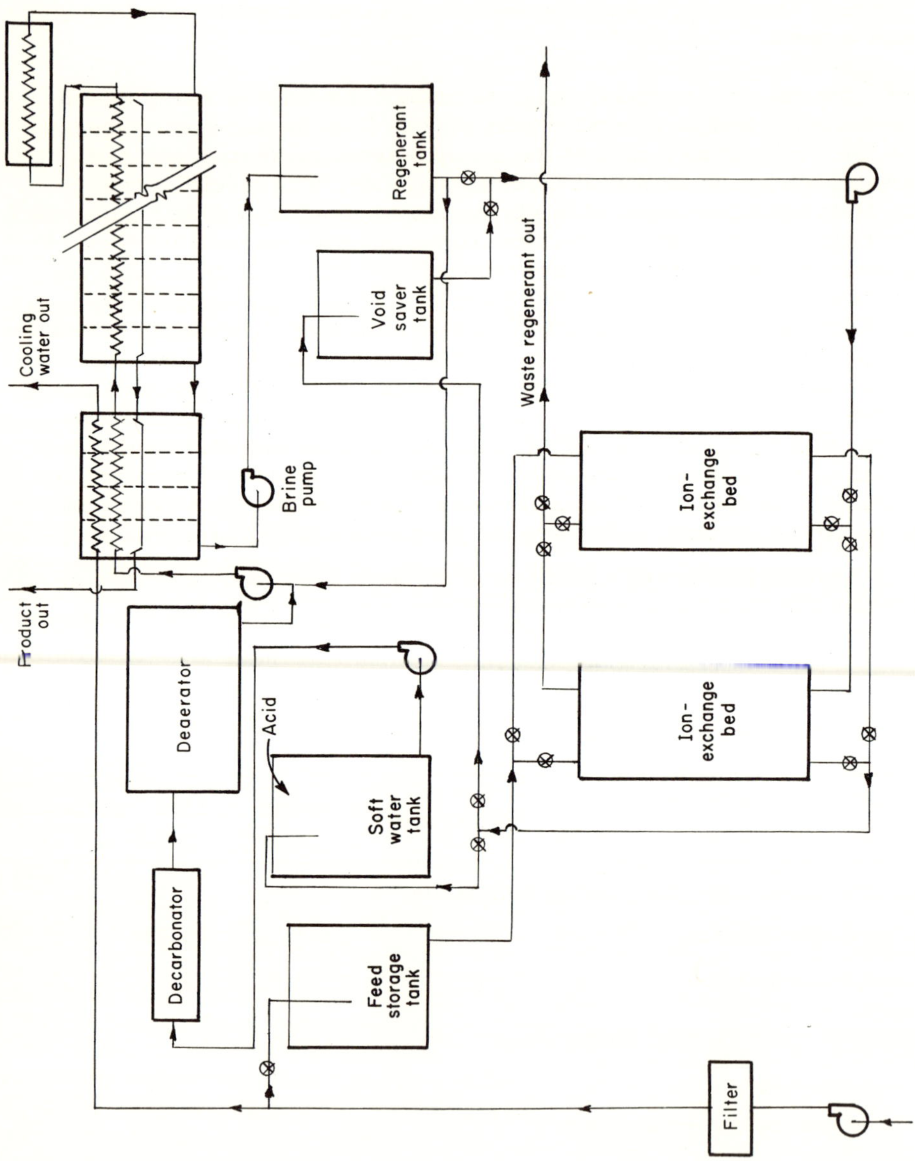

FIG. 5-5. Ion-exchange process for softening seawater feed.

more of the offensive ions. The most obvious scheme would be to use ion exchange to replace either the calcium ions with sodium ions or the sulfate ions with chloride ions. The ion exchanger material could be regenerated by using the concentrated brine effluent from the plant. In a detailed study of this process, directed toward the removal of the calcium ion, Klein et al. [12] estimated that the distiller brine could be concentrated to four times that of normal seawater and that this concentration would furnish sufficient sodium ions to regenerate the ion exchanger. The general arrangement of a plant using this process together with acid treatment for alkaline scale control is shown in Fig. 5-5. While the ion exchange process seems to offer certain attractive possibilities it has not found general acceptance for high temperature plants.

Besides the ion exchange process, the lime-magnesium-carbonate process offers to date perhaps the most effective means of eliminating calcium ions. This is a two-step treatment of the feedwater before it ever enters the distiller plant, and involves first, the addition of magnesium carbonate to precipitate part of the calcium as calcium carbonate. The second step follows with the addition of calcium hydroxide from slaked lime to precipitate more of the calcium as calcium carbonate. This two-step process is shown in Fig. 5-6, and the reactions occurring in the two steps are expressed as follows:

$$CaCl_2 + MgCO_3 \rightarrow MgCl_2 + CaCO_3(ppt)$$

$$Ca(HCO_3)_2 + Ca(OH)_2 \rightarrow 2H_2O + 2CaCO_3(ppt)$$

If these two steps go to completion, both the amount of calcium and the bicarbonate ion remaining in the feedwater before it enters the plant will be very small, thus preventing the formation of both the alkaline scale components and the $CaSO_4$ scales. An installation of this type was installed to furnish softened water to the 1 mgd flash distillation plant at San Diego, and was reported to have removed over 30% of the calcium from the feed with the prospect that process improvements could raise this percentage to at least 65%, the original design objective. With this large amount of calcium removed it would be possible to raise the brine temperature to about 320°F, thus considerably improving the distiller economy.

Recent developments of reverse osmosis membranes have given rise to the suggestion that this process could be used to remove the calcium or sulfate ions from seawater. A study conducted by the University of California at La Jolla [13] has developed a program attempting to produce membranes capable of desalting seawater to potable quality and has accomplished this to the extent of reducing seawater

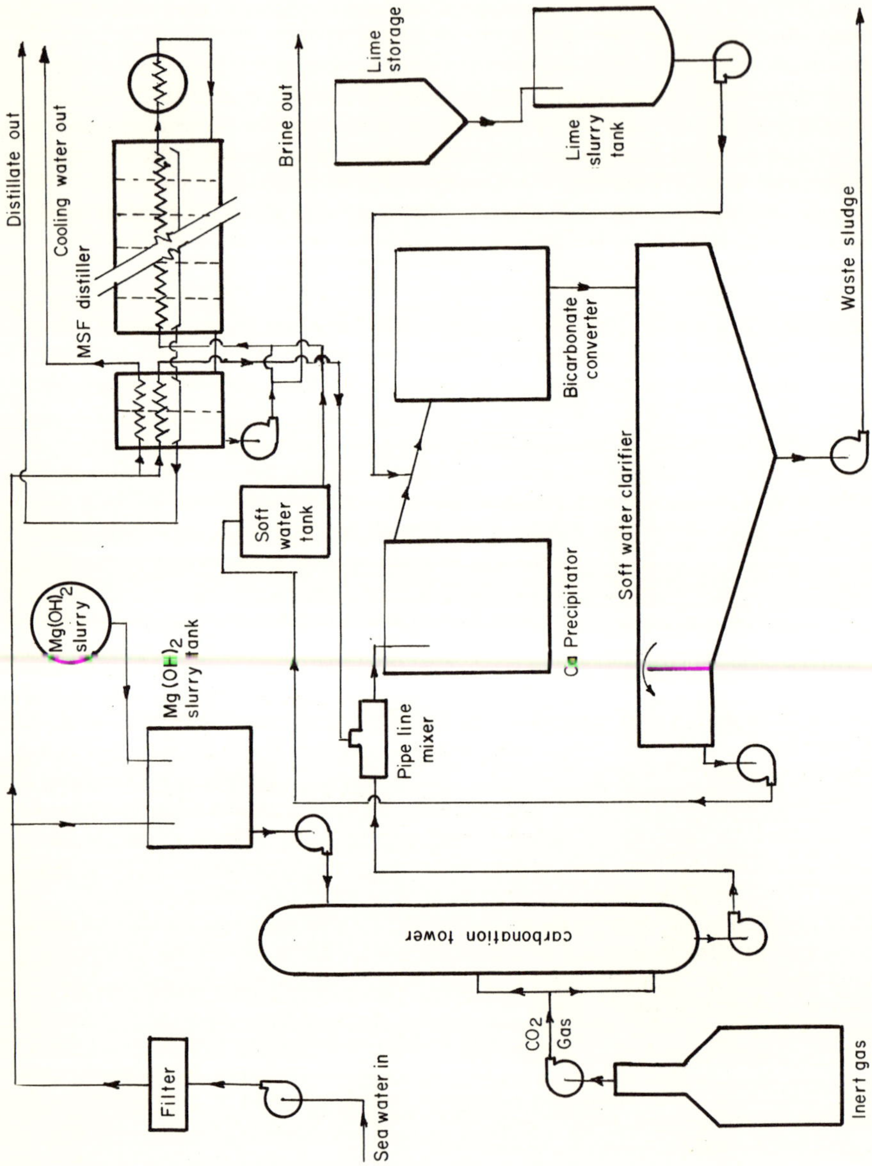

FIG. 5-6. Lime-magnesium-carbonate (LMC) process for softening seawater.

salinity from 36, 231 to 22, 649 ppm. In the study it was discovered that certain reverse osmosis membranes have the selective property of being capable of removing ions having the higher valences. Membranes of several different porosities were tested, using a water pressure of 600 psi, and the most porous of these reduced the sulfate ions by nearly 90%, and the total calcium, magnesium, and bicarbonate ions by over 60%. These results suggest that membranes of greater porosity might prove effective for removing unwanted, high-valence ions from seawater feed at a moderately low cost for equipment requiring the consumption of relatively little energy. At this time it is too early to report any installations using reverse osmosis to prepare seawater for distilling, but the possibilities are encouraging.

In summarizing the results of work on the control of calcium sulfate scale, it should be realized that the operation at most plants is arranged to take place with temperatures at all points in the distillation process lying definitely below the solubility lines for $CaSO_4$ and its hydrates. Ion exchange and the lime-magnesium-carbonate process, with the possible development of reverse osmosis, seem the most likely processes to become economical for softening seawater in order to permit the increase of maximum brine temperatures above 235°F.

B. Deaeration

The presence of dissolved gases in the feedwater presents a condition which has two undesirable effects; they tend to blanket the heat-transfer surfaces and thus interfere with heat flow, and any oxygen present is likely to contribute to corrosion of the heat-transfer tubing. To reduce the hazard from these effects, the feed to a distiller is usually deaerated. Since the feed is separated from the cooling water just after it has passed through the heat-rejection section of the plant, deaeration is initiated usually at a temperature only slightly greater than that of the water source.

The amount of dissolved gases that can be removed by deaeration will be determined by the solubility of the gases and the degree to which the water is saturated with them. In the case of seawater it is usual to assume that the water has become saturated with gases from the atmosphere, due to wave action and other local disturbances. The amounts of oxygen and nitrogen contained in the water can be estimated by means of Henry's Law. These estimates have been shown by Chambers [14] to be in accord with experimental observation. The concentrations of oxygen and nitrogen at saturation are shown in Fig. 5-7 for the range of temperatures from 50 to 120°F, within which the

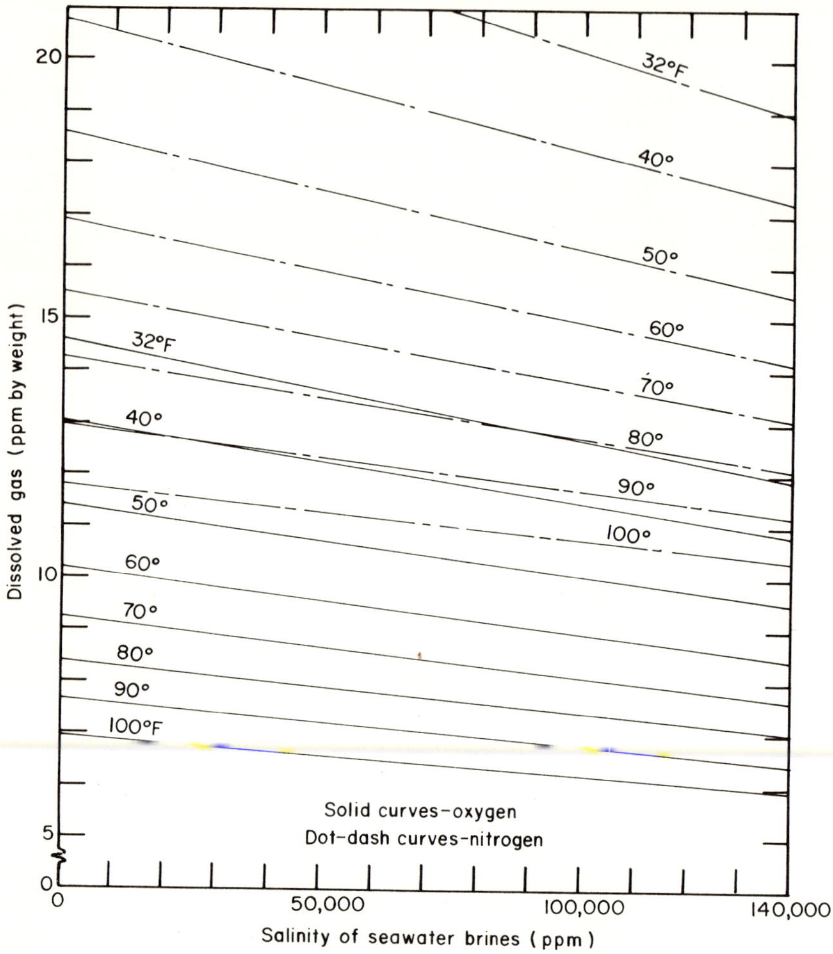

FIG. 5-7. Solubility of gases in seawater. Solid lines, oxygen; dashed lines, nitrogen.

procurement and deaeration of feedwater generally takes place. From this figure it will be seen that the ratio of nitrogen to oxygen is less in solution than in the atmosphere. It is also evident from the curves that the increase in temperature during passage through the heat-rejection section of the plant will have caused some of the dissolved gases to be released from solution. In addition to the oxygen and nitrogen from the atmosphere, any plant using acid for alkaline scale control will have CO_2 gas in the feedwater. The amount of CO_2 will depend on the initial bicarbonate ion concentration. For standard seawater this quantity is about 140 ppm, which would result in the production of approximately 100 ppm of CO_2 upon decomposition. Thus it can be seen that the problem of deaeration consists of the removal of oxygen, nitrogen, and carbon dioxide gases.

The general scheme for effective degassing of liquids is to spread the liquid out into a large surface under saturation conditions in which the solubility of the gases will be zero. The surface area may be either that of a multitude of small drops, as in a spray, or that of a thin film of liquid, as on packing or trays. In some devices the feedwater enters in a spray and then flows through packing material, thereby utilizing both types of surface exposure. In order that the liquid be brought to saturation conditions, some vapor must be made available to exit with the unwanted gases. The necessary vapor may come from an external source or may be produced by flashing of the liquid as it enters the degassing equipment. Several possible arrangements have been developed to accomplish the above treatment of water.

Deaeration of water has been practiced in power plants for many years and has become more or less standardized. In this case, the deaerator, serving as one of the feedwater heaters, is operated at a pressure slightly greater than atmospheric. The water entering near the top of the enclosure is spread over the entire surface of perforated trays or slats and trickles through several layers of these. At the same time, steam at low pressure is introduced at the bottom of the trays and rises countercurrently through the falling water, the steam heating the falling water to a boiling point. Most of this steam condenses so that only a small portion of vapor remains to leave the enclosure with the unwanted gases. Since the pressure in the enclosure is slightly greater than atmospheric, the gases can be discharged without pumping. Although this scheme contains the essential features of the deaeration process for distillers it differs in two essential respects, namely, the size of equipment involved and the temperature of operation. The amount of water that must be treated in even a small distiller plant is considerably greater than that which is treated in a large power plant. Also, the temperature at which deaeration is carried out in a distiller plant is only about 90°F as compared to 212°F in a power plant. These differences are sufficiently important that it has led to the use of new designs of equipment for distiller plants.

Because of the presence of large quantities of dissolved CO_2 in pretreated feedwater, it has been suggested that degassing might be effectively accomplished in two steps. Several plants have been so designed. The first step consists of a tower much like a tower for cooling water in which the feedwater is introduced at the top and trickles over trays or slat packing while air is pumped to flow upward through the falling stream of water. This air serves to strip the liquid of CO_2 at atmospheric pressure, resulting in the reduction of the value of CO_2 to an order of about 10 ppm. With this residue of CO_2 and a saturation load of oxygen and nitrogen, the feedwater then passes to the second step, the vacuum deaerator, in which final degassing is accomplished by steam under vacuum conditions at relatively low temperatures. There is some argument as to whether the atmospheric degassing of the CO_2 followed by vacuum deaeration offers any savings over the method of accomplishing the entire degassing operation in a single vacuum deaerator, even though the latter must be operated at a very high loading.

Most of the deaerators designed for distillers are of the packed tower type, using packing such as Rashig rings. The steam for stripping may be drawn either from an appropriate stage of the distiller or may be produced by flashing of the feedwater itself. Another variation is the direction of the stripping vapor with respect to the feedwater flow. If external steam is used, the flow can be either co- or countercurrent with respect to the water flow, but if the steam is produced by flashing of the feedwater the flow will be cocurrent with respect to the water direction. Although there are no really standard designs for distiller plant deaerators, there seems to be general acceptance of flow rates through these devices, amounting approximately to 20, 000 $lb/(hr \times ft^2)$ for water and 15 to 20 $lb/(hr \times ft^2)$ for stripping vapor. Hammond's [15] suggested standards for feedwater purity are 1 ppm for CO_2 and less than 10 ppb (part per billion) for oxygen. Purity of this order of magnitude has been achieved in existing deaerators.

III. ENTRAINMENT SEPARATORS

After the feedwater passes into the plant one of the difficulties common to all distillers is the prevention of the carryover of droplets of saline water into the condensing section of the plant. This occurs because the evaporation process is a very turbulent one and produces a mist or fog of water drops which tend to be carried along with the vapor due to the viscosity of the vapor and its motion. These drops can collect in the layer of condensate on the cooling surface of the condenser, thereby contaminating the condensate. To prevent the carryover of these droplets it is standard procedure to include a porous entrainment barrier in the path of the vapor between the evaporating zone and the condensing surface.

Various designs of entrainment separators have been used in distiller plants, the general arrangement being to divide the vapor into many streams with each stream following a tortuous path with many changes in direction. At each change in direction the droplets tend to be thrown out of the stream due to centrifugal force, and impinge on any solid surface with which they come into contact. If the solid surface is wettable, the drops will form a film, flow down to the bottom of the solid material, and fall off in drops too large to be entrained by the vapor.

An example of an early design of separator is the hook and vane type in which the vapor flowed horizontally through several rows of vertical sheet metal slats. These slats were formed into U and J shapes, and staggered in the several rows so that the stream passing between vanes in one row would impinge on the concave part of the hook or vane in the next row. Droplets would collect in this concave area, drop to the bottom of the vane, and drain through a hole into the unevaporated saline water. These early separators were expensive and gave rise to an appreciable pressure drop.

The type of separators presently used in most plants consists of a mat made up of many layers of wire mesh, each layer being staggered relative to the next. These mats are placed horizontally and the stream of vapor rises vertically through them. As the vapor rises, the droplets collect on the mesh wires, merge into larger drops, and drip from the bottom layer, the drop size being too large to be entrained by the rising vapor. Separators of this design present very little resistance to vapor flow and are capable of ensuring condensate of very low salinity, usually between zero and about 5 ppm.

IV. POSTTREATMENT OF DISTILLED WATER

As indicated in Chapter 2, distilled water has a tendency to dissolve most materials with which it comes into contact. Because of this, distilled water tends to attack metal pipes or other conduit materials through which it may be channeled for distribution to the consumer. This tendency may be reduced when a relatively small amount of dissolved material is present. Rather than to have this dissolved material come from the piping, which is equivalent to corrosion of the piping system, most distillation plants employ some means of adding at least 50 ppm of dissolved salts to the distilled product. Various methods are used, such as blending of the distillate with a small amount of brackish ground water, the trickling of distilled water through a filter bed made of crushed coral rock, or adding a small quantity of lime to the distillate. In the case of the blending with a supply of slightly saline water, the salinity may be raised considerably above the 50 ppm level required to prevent corrosion, thereby "stretching" the quantity of freshwater being made available. For example, blending distilled water of 5 ppm salinity with about 17% of saline water of 2000 ppm salinity would result in a product of 300 ppm salinity.

REFERENCES

1. Stearns-Roger Mfg. Co., 2nd Annual Rep., Saline Water Conversion Demonstration Plant, Freeport, Texas, O.S.W. Res. Develop. Progr. Rep. No. 100, 1961, 84 pp.
2. Condensate from Mine Plant Supplies Town with Water, <u>Coal Age</u>, <u>34</u> (No. 5), 1929, p. 292.
3. G.E. Sonderman, Sea Water Distillation with By-product Power at Aruba, <u>ASME Paper</u>, <u>No. 59-SA-24</u>, Mar. 1959.
4. Bechtel Corp., Engineering and Economic Feasibility Study, Phases I and II, for a Combination Nuclear Power and Desalting Plant, U.S. AEC Div. of Technical Information, Dec., 1965, 13 Chapt.
5. M. N. Elliot, The Present State of Scale Control in Sea Water Evaporators, <u>Desalination</u>, <u>6</u> (No. 1), 1969, pp. 87-104.
6. D.F. Othmer, Evaporation for Desalination - Scale Prevention and Removal, <u>Desalination</u>, <u>1</u> (No. 2), 1966, pp. 194-198.
7. V.B. Chernozubov, et al., Prevention of Scale Formation in Distillation Plants by Means of Seeding, <u>Desalination</u>, <u>1</u> (No. 1), 1966, pp. 50-60.
8. C.H. Lu and B.M. Fabuss, Calcium Sulfate Scaling in Saline Water Distillation, <u>Ind. Eng. Chem. Process Design Develop.</u> <u>7</u> (No. 2), Apr., 1968, pp. 206-212.
9. W.L. Marshall and R. Slusher, Solubility of Calcium Sulfate and its Hydrates in Sea Water and Saline Water, Paper presented at ACS meeting, 1967.
10. J.W. McCutchan and E.N. Sieder, Scale Control in Saline Water Evaporators - a Review of Current Status, Res. Develop. Progr. Rep., No. 411, Office of Saline Water, Apr., 1969, 62 pp.
11. Dow Chemical Co., An Engineering Evaluation of the Long-Tube Vertical Falling-Film Distillation Process, Res. Develop. Prog. Rep., No. 139, Office of Saline Water, Feb., 1965, 282 pp.
12. G. Klein, S. Cherney, E.L. Ruddick, and T. Vermeulen, Calcium Removal from Sea Water by Fixed Bed Ion Exchange, <u>Desalination</u>, <u>4</u> (No. 2), 1968, pp. 158-166.

13. J. W. McCutchan et al., Saline Water Demineralization by Means of a Semipermeable Membrane, Progress Summary, Saline Water Research, WRC Desalination Rep., No. 47, Univ. of Calif., Los Angeles, California, Jan., 1972, pp. 1-12.
14. J. T. Chambers, Performance Characteristics of Vacuum Deaeration Equipment, Univ. of Calif., Berkeley, Inst. of Eng. Res. Issue No. 16, Series No. 75, Sept. 3, 1959, 27 pp.
15. P. Hammond, Deaerators for Desalination Plants, O. S. W. Res. Develop. Progr. Rep. No. 314, Office of Saline Water, Dec., 1967, 117 pp.
16. L. S. Herbert and U. J. Sterns, Saline Water Distillation – Scaling Experiments Using the Spray Evaporation Technique, *Advan. Chem. Ser.*, *38*, 52 (1963).

Chapter 6

MULTIPLE-EFFECT DISTILLATION

I. DEVELOPMENT OF THE MULTIPLE-EFFECT DISTILLATION PROCESS

Multiple-effect distillation was conceived as a direct means of reusing heat energy, and thereby increasing the yield of distilled water from a given amount of heat input above that possible with a single-effect plant. The basic principles involved have been explained in Chapter 4, so that the purpose of this chapter is to consider the manner in which these principles are actually applied to the production of distilled water. Multiple-effect distillers have been used over a longer period than any other type, probably beginning with the early days of steam propulsion for ships. Modern applications are numerous, and are prominent in salt recovery and sugar production as well as in water purification. A partial list of noteworthy water distillation plants using this scheme is shown in Table 6-1. This list includes only plants which are used primarily for potable water supplies and does not contain the many units constructed to supply distilled water for boiler feed makeup. It should be noted that the more inclusive OSW inventory [1] has listed 293 multiple-effect plants, classified as the submerged-tube type, and 92 plants of the long-tube vertical (LTV) and vertical tube evaporator (VTE) types.

TABLE 6-1

Representative Multiple-Effect Distillation Plants

Location	Type[a]	GOR	Number of units	Capacity (in 1000-gpd's) Per unit	Total	Year
Aden, Hedjuff	ST		2	17	34	1962
Aruba, Govt Water Works	ST	4.6	5	500	2688	1958
Curacao, Mundo Nobo 7-8	ST	4.6	2	500	1074	1959
Curacao, Mundo Nobo 1-6	ST		6	80	483	1948
Gibraltar	ST		2	80	161	1947
Kuwait, Shuwaikh	ST	2.3	10	125	1260	1956
Kuwait, Shuwaikh	ST	2.3	10	135	1344	1956
Kuwait, Mina-al-ahmadi	ST	2.3	8	120	960	1954
Netherlands, Pernis	VTE		1	700	700	1966
Netherlands, Leiden	VTE		1	177	177	1966
Turkey, Istanbul	ST		1	129	129	1952
U.S., Freeport, Texas	LTV	10.5	1	1000	1000	1961
Virgin Islands	VTE		1	1000	1000	1968

[a] ST, submerged tube design; VTE, vertical tube evaporator design; LTV, long-tube vertical design.

Most of these plants in the OSW list were built to serve power plants and, while these installations have contributed substantially to the development and evolution of multiple-effect distillation for potable usage, the emphases in such plants have been more on high purity of the product and trouble-free operation than on economy of large-scale water production. Steam economy, which is of paramount importance in plants producing large quantities of potable water, is of negligible importance for producing boiler feed in power plants, inasmuch as the distillation unit merely constitutes a minor section in the feedwater heating system, and the net heat required for water production is only that which is discharged as sensible heat in the blowdown brine from the distillation unit. This is the chief reason that the plants merely supplying boiler feedwater were omitted from Table 6-1.

II. DESIGN CONFIGURATIONS

In Table 6-1 the plants are designated as being of several different types. These reflect the application of various design configurations, the best known of which are (1) the submerged tube design (ST), (2) the long-tube vertical design (LTV), and (3) the vertical tube evaporator design (VTE).

The submerged tube concept, as the name implies, is that of a heating coil submerged in a tank of saline water, with steam condensing inside the coil and vapor being expelled from the surface of the water outside the coil. Very little attention is given to preheating the incoming water in this concept because the emphasis is on the transfer of heat from condensing steam to the evaporating water. The early devices using this concept had flat spiral or helical coils arranged to promote the circulation of the water being evaporated. As sizes of plants were increased these early coils were replaced by tube bundles placed horizontally, and extending the full length of horizontal cylindrical shells, as shown in Fig. 6-1. The largest plant built, using this configuration, was the Aruba plant, constructed in 1958, having five 6-effect distillation units, each capable of producing approximately 500, 000 gpd. No large plants of this type have been constructed since that time, since the development of the multiple-stage flash distillation design has apparently taken preference.

The major problem of the submerged tube configuration was the lack of positive circulation of the saline water and the resulting tendency for scale to form on heating surfaces where brine velocities

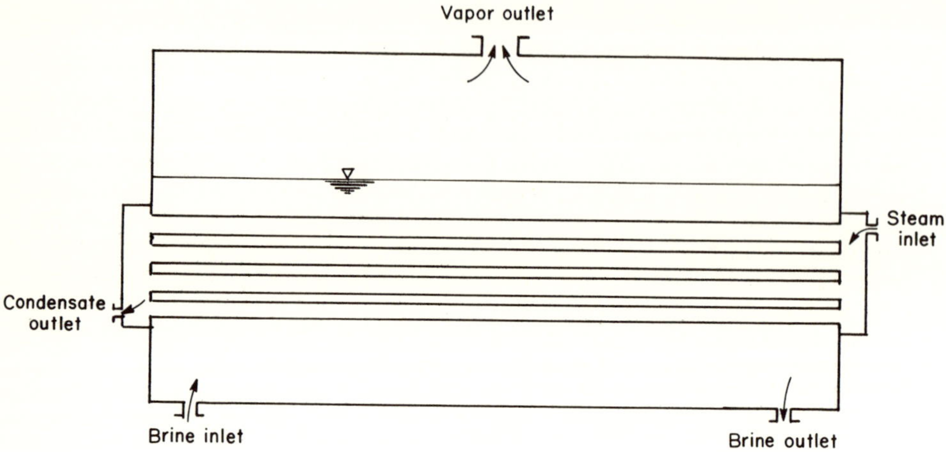

FIG. 6-1. Submerged tube evaporator. (Schematic only.)

were low. The long-tube vertical (LTV) concept was advanced as a design using positive circulation, a configuration which had been proved in the salt industry and had been found to be relatively free from scaling problems. The concept in this case was that of a thin sheet of saline water falling vertically in contact with a metallic surface heated by condensation of steam on the opposite side of the surface. In order to pack a large area of vertical surface for evaporation into minimum space, the surface used was the inner surface of tubes running vertically between tube sheets approximately 20 ft apart, as shown schematically in Fig. 6-2. This concept was used in the first demonstration plant of the OSW, a 12-effect plant constructed at Freeport, Texas and capable of delivering 1 mgd of distilled water. In addition to the LTV evaporator-condensers, the Freeport plant included an elaborate system of feedwater heaters, part of these being supplied with steam from the several effects and part of them liquid-to-liquid coolers for the condensate. In addition, the plant was laid out with pumps for the removal of brine from one effect and delivery to the next. The Freeport plant was operated for over 10 years, and during the course of this time was used for extensive experimentation, which was instrumental to the development of the VTE concept.

The VTE concept, an abbreviation for vertical tube evaporator, is similar to the LTV concept in that it is basically one in which vapor forms from a thin film of water flowing by gravity down a vertical metallic surface, with the latter heated by vapor from the preceding effect condensing on the opposite side of the surface. The surface is

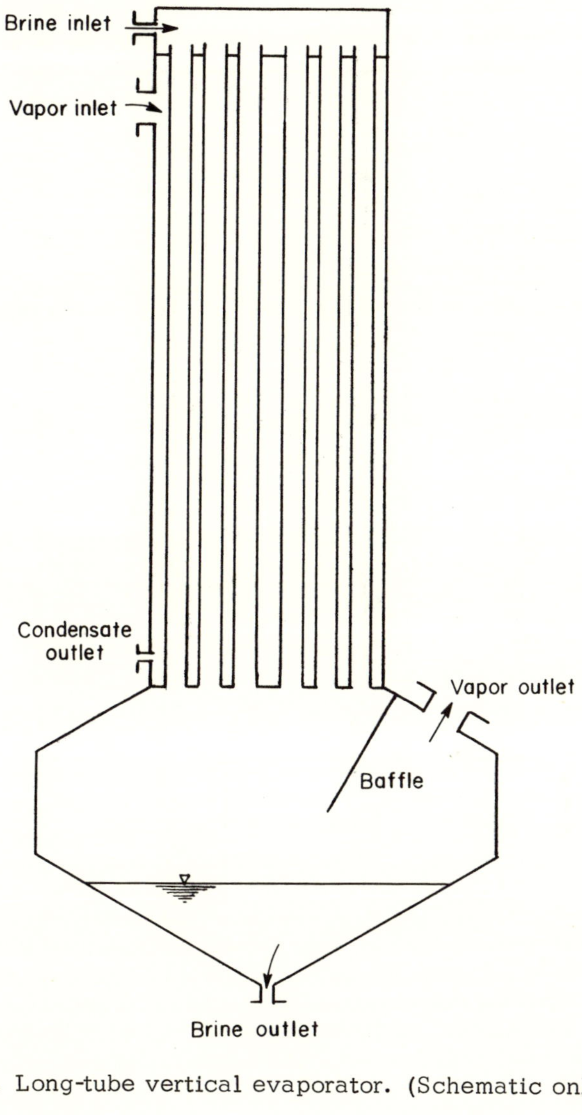

FIG. 6-2. Long-tube vertical evaporator. (Schematic only.)

in the form of vertical tubes, as in the LTV, but the tubes in the VTE concept are relatively short and are generally enhanced from the viewpoint of heat transfer. From experience derived from the Freeport plant, the pumping of brine from one effect to the next is eliminated and is replaced with the passage of the brine from effect to effect by means of the pressure gradient. Pumps are used, however, to lift the brine

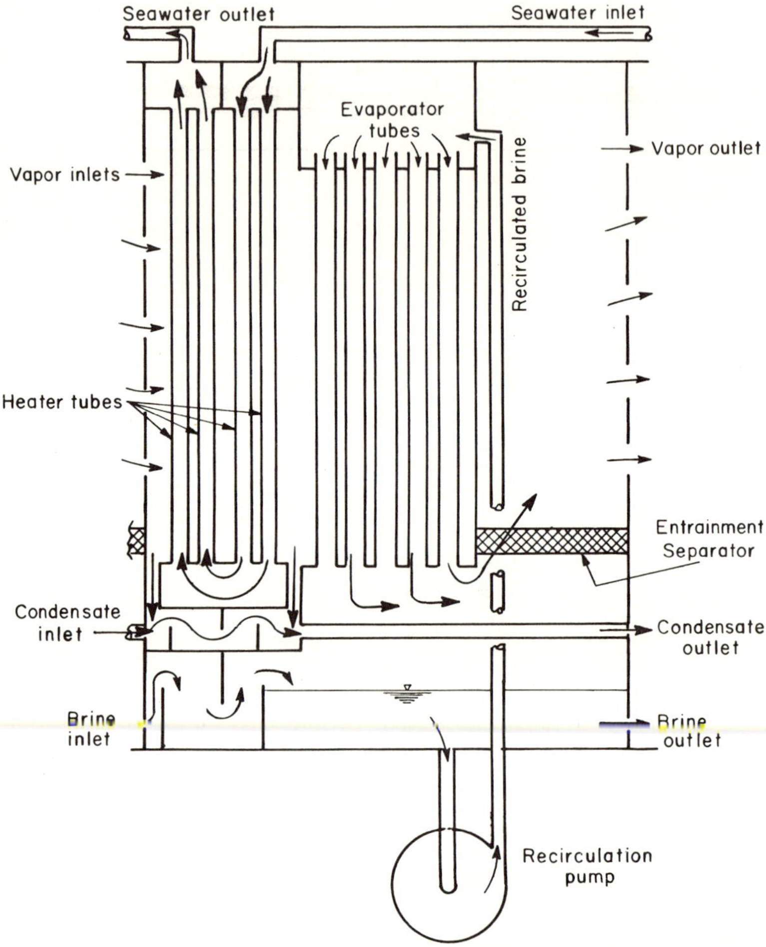

FIG. 6-3. Vertical tube evaporator module. (Schematic only.) Note that module includes feedwater heater element.

from the lower sump and deliver it to the top of the tubes in the same effect. The progress of the condensate from effect to effect by means of the pressure gradient is accompanied by flashing which thereby cools the condensate and eliminates the need for the liquid-to-liquid coolers of the LTV plant. The general arrangement of a VTE evaporator is shown schematically in Fig. 6-3. This figure implies a modular arrangement, i.e., one in which a single shell or case encloses several similar effects. Such an arrangement eliminates much of the external piping found in older plants and is particularly favorable to the development of plants of large capacity. Many of these features have been incorporated in the design of the 1 mgd plant constructed at Christiansted in the Virgin Islands [2]. While all three of the concepts are found in existing plants, the VTE configuration appears to be the most favored for the very large plants which may be constructed in the future.

III. SIGNIFICANCE OF THE NUMBER OF EFFECTS

Besides showing different types of multiple-effect distillers, Table 6-1 also indicates great ranges of size and economy among them. It will be observed that the per unit capacity in some plants is as large as 1 mgd and that the GOR value varies from 2.3 to approximately 10. In Chapter 4 it has been shown that the GOR of a plant is a measure of its economy, and in a multiple-effect plant is proportional to the number of effects; in other words, the larger the number of effects, the greater the thermal efficiency. Because of this relationship to steam economy, it is important to take into account some primary conditions which must be considered in the selection of the number of effects. This is not a simple decision because, as will be seen in Section III, B, the number of effects shows little correlation with capacity.

A. "Ideal" Maximum Economy

Regardless of the plant design configuration, the selection of the number of effects for a particular plant is a prime factor in determining steam economy and consequent water costs. Since the GOR is a measurement of steam economy and is directly proportional to the number of effects in a multiple-effect plant, it is important to inquire as to the maximum number of effects which could be used for any given temperature range.

The maximum number of effects would be the number that would permit heat transfer to occur with the smallest possible difference in temperature in each effect. This would be an infinitesimal temperature difference. Heat, transferred at this small temperature difference, would be used in forming vapor and thereby supplying heat to the next effect by its condensation. This vapor condenses at a temperature less than that at which it was formed, by the amount of the boiling point elevation (BPE), so that each successive effect will be lower in temperature than the preceding one by at least the amount of the BPE. Although the BPE varies from one effect to the next because of the change in salinity, an average value of about 1 °F is the approximate BPE encountered in the distillation of seawater. Therefore, an "ideal" distiller of the multiple-effect type would have a number of effects approximately equal to the operating temperature range of the plant. For example, in a plant operating between the temperature limits of 220 and 100 °F, the ultimate limit of number of effects that it would be theoretically possible to include in that temperature range would be 120 effects [(220 - 100)/1.0 °F]. A GOR, corresponding to that number of effects and based on a 0.8 factor, the logic of which will be explained in more specific detail later in this chapter, would be 96 and represents the absolute "ideal" maximum GOR achievable under "ideal" conditions of infinitely large heat-transfer surfaces in each effect. This value obviously exceeds the conditions found in any actual plant. For instance, the "ideal" number of effects and GOR of the above illustration is nearly ten times larger than the best of any accomplished performance, notably, the 12-effect plant at Freeport, Texas. As of the present date, the largest number of effects operated experimentally is 30, realized by Bromley [3] in a rotating device with flat-plate heat-transfer surfaces. The GOR was in excess of 20.

From the above arguments it should be clear that the boiling point elevation of the water being evaporated imposes the effective limit to the number of effects possible in an "ideal" multiple-effect distiller. This limit is far larger than can be realized in any actual plant and is only of interest because it sets the absolute limit within a given temperature range.

B. Relation between Plant Capacity and Number of Effects

The preceding section has shown that, although the "ideal" number of effects is proportionately dependent upon the available temperature range, the actual, practical number of effects in any existing plant is far less than the "ideal" and has little relationship to the temperature range, but rather to steam economy. In contrast, the capacity of a plant bears little relationship to the number of effects, but does have a direct relation to the temperature range, as will be seen in this section.

That the number of effects has practically no influence on the capacity of a plant will be understood as one considers the heat flow problem in the evaporator-condensers. In a multiple-effect plant the difference between the condensing temperature of the steam supply and the temperature of the incoming cooling water is divided among the several effects, so that each has about the same differential temperature to accomplish heat transfer.

To simplify the reasoning of this concept, one can visualize a plant consisting of n evaporator-condensers, each having the same heat-transfer area and so arranged that the overall heat-transfer coefficients are equal in all effects. Ignoring the change in latent heat at different temperatures and the change in boiling point elevation, the capacity of the plant will be proportional to the total amount of heat transferred from condensing vapor to evaporating water. The magnitude of heat flows will be as follows:

First effect $$q_1 = UA(\theta_m)_1 \tag{1}$$

where q is rate of heat flow, U is overall heat-transfer coefficient, A is heat-transfer area in each effect, and θ_m is the temperature difference causing heat transfer.

Second effect $$q_2 = UA(\theta_m)_2 \tag{2}$$

nth effect $$q_n = UA(\theta_m)_n \tag{3}$$

The total amount of heat used in forming the distillate will be the summation of the above, so that

$$(q_1 + q_2 + \cdots + q_n) = UA[(\theta_m)_1 + (\theta_m)_2 + \cdots + (\theta_m)_n] \tag{4}$$

but the bracket on the right-hand side is the major share of the total temperature drop in the apparatus, and can be written as follows:

$$t_h - t_c = [(\theta_m)_1 + (\theta_m)_2 + \cdots + (\theta_m)_n] + (\theta_m)_{n+1} \tag{5}$$

where t_h is condensing temperature of steam, t_c is the inlet temperature of cooling water, and $(\theta_m)_{n+1}$ is the temperature difference causing heat transfer in final condenser.

If it is assumed that the values of θ_m are equal in all effects and in the final condenser, Eqs. (4) and (5) can be combined to give:

$$\sum q = UA(t_h - t_c)[n/(n+1)] \tag{6}$$

In Eq. (6) the term Σq is proportional to the amount of distillate produced, and the right-hand member of the equation indicates that this same amount of water could be produced by a single-effect plant provided the entire temperature difference were employed to cause heat transfer. From the above reasoning it is deduced, therefore, that a given plant capacity can be achieved with any number of effects, and that capacity is essentially independent of the number of effects, but rather is proportional to the temperature range, $t_h - t_c$.

C. Economic Optimization of the Number of Effects

The foregoing paragraphs set forth the desirability of many effects from the viewpoint of steam economy in spite of the fact that the number of effects has little influence on capacity. However, in opposition to the steam economy advantage is the need to minimize this number from the viewpoint of capital cost. For any given situation the compromise between these two extremes must be found by optimizing the number of effects which will give minimum cost for the product water. While true optimization should involve precise values for many variables at different points in the plant, some significant general conclusions can be reached by resorting to the use of average values in appropriate equations.

As has been pointed out the chief costs which must be considered in the operation of a multiple-effect distillation plant are capital-amortization and steam costs. For simplicity, capacity is considered a constant in this analysis. One approach to determining these cost elements can be formulated by the use of the scheme described by Chambers [4]. To use his formulation it is assumed that the saline water is supplied to the first effect at its boiling point, that all effects and the final condenser have equal heat-transfer areas and equal heat-transfer coefficients, and that all effects have equal distillate production rates. This situation is represented graphically in Fig. 6-4.

In each effect the quantity of heat transferred will be

$$q = \Delta W \cdot h_{fg} = UA_e \theta_m \tag{7}$$

where ΔW is distillate produced, h_{fg} is latent heat of distillate, U is overall heat-transfer coefficient, θ_m is mean temperature difference for heat transfer, and A_e is heat transfer area of one effect.

The total production of the plant can then be written as

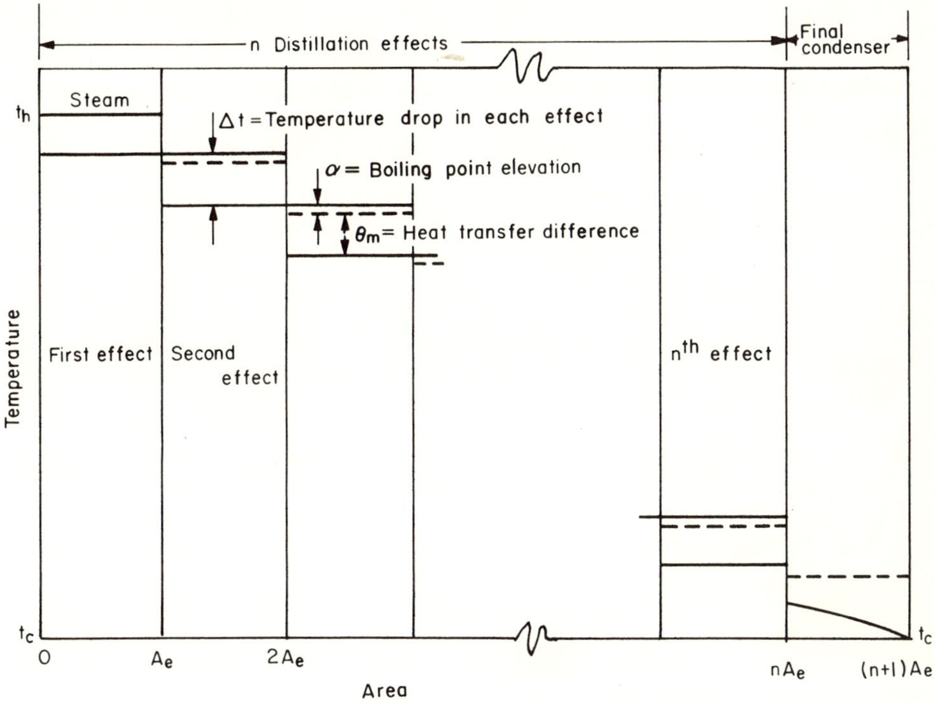

FIG. 6-4. Temperature area diagram for a multiple-effect distiller, having n effects.

$$W = \sum \Delta W = (n/h_{fg})UA\theta_m \tag{8}$$

However, since $\theta_m = \Delta t - \alpha$ where α is the boiling point elevation,

$$W = (UA/h_{fg})(n\Delta t - n\alpha) \tag{9}$$

Since there is no boiling point elevation in the first effect, the total temperature difference may be written as:

$$\Delta T = (t_h - t_c) = (n+1)\Delta t - \alpha \tag{10}$$

From this

$$n\Delta t = [n/(n+1)]\Delta T + [n/(n+1)]\alpha$$

Noting that the entire heat-transfer area of the plant may be written:

$$A = (n + 1)A_e$$

the final form of Eq. (9) is:

$$W/A = (U/h_{fg})[n/(n+1)^2](\Delta T - n\alpha) \tag{11}$$

Equation (11) shows that the value of W/A decreases as the number of effects is increased. This means that a plant of a given capacity, W, would need greater heat-transfer area, A, for the larger number of effects. Since the capital cost of the plant increases in approximate proportion to the area, A, the larger number of effects corresponds to a larger capital cost. At the same time, any increase in the number of effects produces an increase in the steam economy. Thus, as the number of effects increases, the capital cost increases and the cost of steam decreases, leading to the conclusion that there should be some particular number of effects for which the sum of the capital cost and the heat energy cost will be a minimum.

The contributions of the energy cost and capital cost to the unit cost of water, per thousand gallons, can be written in terms of the steam economy and the area economy as indicated in the following equation:

$$C^* = \frac{0.9513\,ap}{W/A} + \frac{8333bh_{fg}}{W/W_s} \tag{12}$$

where C^* is cost of water (dollars/lb), a is capital cost of plant per ft^2 of total heat-transfer surface, p is annual charges on capital, as a fraction, b is cost of steam in dollars per million Btu, W is distillate production rate in lb/hr, W_s is steam supply rate in lb/hr, A is area of heat-transfer surface in ft^2, 8333 is pounds of water per kgal, and 0.9513 is ratio of 8333 to hours in a year.

To show the result of changing the number of effects in the plant, let $W/W_s = 0.8\,n$ and substitute the right-hand side of Eq. (11) for W/A so that Eq. (12) becomes:

$$C^* = \frac{0.9513\,ap}{(U/h_{fg})[n/(n+1)^2](\Delta T - n\alpha)} + \frac{8333\,h_{fg}b}{0.8\,n} \tag{13}$$

This equation has the general form

$$C^* = \frac{X(n+1)^2/n}{\Delta T - n\alpha} + \frac{Y}{n} \tag{14}$$

where X and Y are numbers representing the constants in Eq. (13) for any given set of conditions. Equation (14) is now in a form which can be differentiated to find the minimum value, the results of such differentiation being as follows:

$$(n+1)(n-1) + [n\alpha(n+1)^2]/(\Delta T - n\alpha) = (Y/X)(\Delta T - n\alpha) \qquad (15)$$

To illustrate the application of the above equations consider a case in which the magnitudes of the basic variables are as follows:

$\Delta T = t_h - t_c = 140\,°F$

$\alpha = 1.0\,°F$

b = \$0.50 per million Btu

p = 14% per annum (interest plus sinking fund)

a = \$6.00 per ft^2 of heat-transfer surface

U = 600 Btu/(°F × ft^2 × hr)

h_{fg} = latent heat of steam, 966 Btu/lb for external steam and 1020 Btu/lb for distillate

These values give X = 1.362 and Y = 5.03, so that the solution of Eq. (15) results in a value of nearly 20 effects, as being the number that would produce the minimum sum of the costs. Using 20 effects in Eqs. (13) and (14), the sum of the two cost components is \$0.50 per kgal.

To illustrate the variation of the cost components with changes in the number of effects, values of each of the terms on the right-hand side of Eq. (13) were computed for several different numbers of effects and plotted on Fig. 6-5, together with a curve for their sums. From this figure it will be seen that the increase in capital amortization is nearly linear with the increase in the number of effects but that the cost of steam decreases at a diminishing rate with the increase in the number of effects. As a result of this difference in the slopes of the two curves, the third curve, representing the sum of the two cost components, is rather flat in the vicinity of the minimum point. Thus, in the case cited, an increase from 10 to 15 effects would decrease the cost of water by nearly \$0.11 per kgal, but a further increase to 20 effects would decrease the cost by only an additional

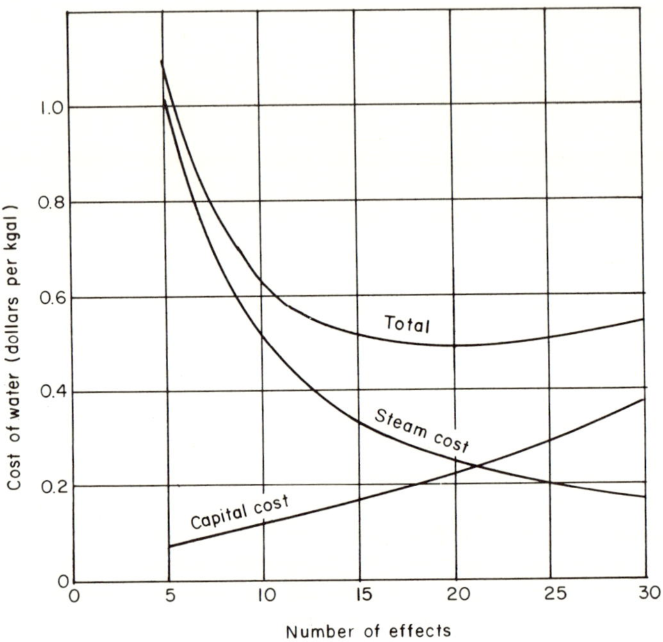

FIG. 6-5. Variation of water cost components with number of effects.

\$0.02. It is also observed from the curves that minimum total cost occurs at the point where the steam and capital cost components are about equal.

Different numbers of effects and minimum costs would result if other values were inserted in Eq. (13). From an inspection of the equation and Fig. 6-5, it is clear that raising the steam cost above \$0.50 per million Btu would require an increase in the number of effects in order to optimize for minimum cost, whereas increasing the capital charges would necessitate a decrease in the number of effects. It is recognized that these calculations, based on average values, are limited in their applicability. However, they do show a first approximation for the number of effects which should be considered for a plant, prior to an opportunity to make more precise estimations when exact design information may become available.

IV. PLANT ARRANGEMENTS

A. Arrangements of Evaporator-Condensers

The above comments concerning the selection of the design configuration and the number of effects have dealt mainly with the inherent potentiality of the heat exchangers to accomplish evaporation and condensation of the water. This potentiality can also vary with the arrangement of the several heat exchangers in a plant and with the course chosen to channel the water through the effects and the auxiliary equipment required for the heat exchangers to serve their stated functions. One of the first decisions in the selection of the design concept is to determine whether the heat exchangers should be operated in a horizontal or vertical configuration. Next it must be decided whether the plant is to consist of several discrete heat exchangers connected with external piping or is to be enclosed in a single casing, with internal passages for conveying the brine, vapor, and distillate from one effect to the following. While most of the existing plants are of the externally piped variety, there has been increasing interest in the integral-type plant for the very large sizes which may be built in the future. Regardless of the question of internal passages or external piping, the designer must provide for the conveyance of the several fluids between the effects, through the use of passages, pumps, and control devices. A very simple arrangement of the most essential auxiliary devices required to move the various streams of water and vapor through a three-effect plant is shown schematically in Fig. 6-6.

B. Forward Feed and Backward Feed

In Fig. 6-6, the three evaporator-condensers are numbered in the order of decreasing temperature, and this is the order in which the vapor and condensate progress through the plant. It is possible, however, to route the seawater feed either in the same direction, called forward feed, or to pipe the seawater feed to enter the plant at the last effect and to progress through the evaporator-condensers in a direction opposite to the flow of condensate. This scheme is known as backward feed.

The plant arrangement of Fig. 6-6 exemplifies forward feed of the brine, i. e., the seawater at its highest temperature enters the evaporating part of the plant cycle at the first effect, flows through the

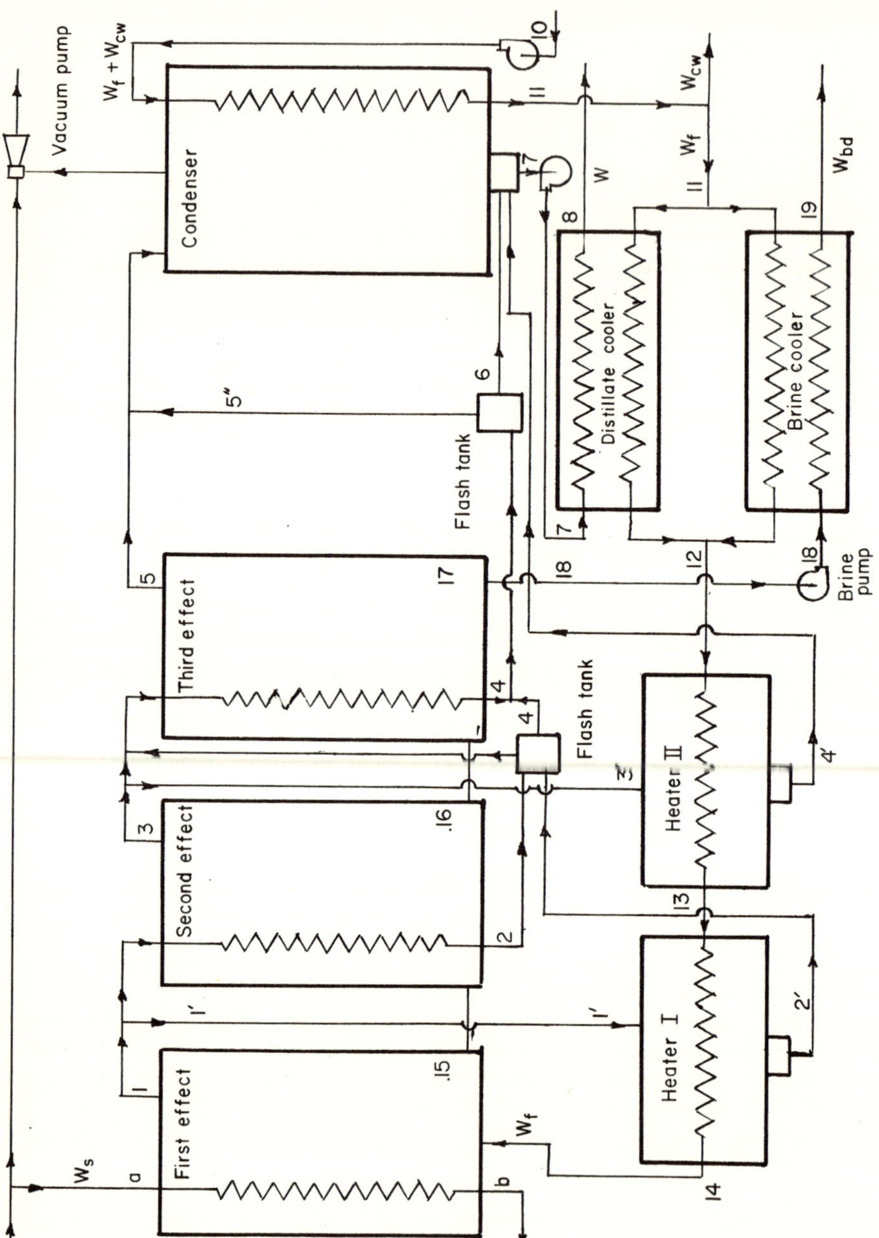

FIG. 6-6. Triple-effect distiller plant. (Schematic only.)

second and third effect in that order, and leaves the plant from the third, or low-temperature, effect. Backward feed would be an arrangement in which the seawater at relatively low temperature would enter the evaporating section of the plant at the third, or last, effect, from which it would be conveyed sequentially through the rest of the evaporating section in reverse order, i. e., to effects of increasing temperature. In this latter arrangement the brine is heated by the condensing vapor of the higher temperature effects to which it is progressing; in turn, this heated brine releases vapor which will be condensed in the lower temperature effects from which it has just come. In this manner the backward feed flow causes the saline water and its brine to be heated to a new boiling temperature in each successive effect, and it leaves the plant at its highest temperature.

This procedure results in slightly better steam economy than the operation of a plant arranged for forward feed, as in Fig. 6-6. The improvement is due to the fact that in forward feed all of the seawater feed must be heated to the maximum temperature, whereas in backward feed only the residue of brine left after vapor formation in the low-temperature effects must be raised to the maximum temperature of the first effect. This superior economy, together with the elimination of separate feedwater heaters, has led to a preference for backward feed in nonscaling evaporation processes, such as in sugar refining and salt production from brines. However, in backward feed maximum brine density occurs at maximum temperature and thus tends to promote scale formation in the first, or high-temperature effect when there are scale-forming ions present, as is the case in most saline waters used for desalting. Other disadvantages of backward feed include the requirement of additional pumps to boost the brine pressure sufficiently to overcome the higher saturation pressures of the high-temperature effects, and the problem of brine rejection at the temperature of the first effect. For these reasons, even though backward feed is admittedly superior for certain chemical processes, forward feed is used almost universally for the production of distilled water.

C. Auxiliary Equipment

From previous comments it is evident that multiple-effect plants require many sorts of equipment besides evaporator-condensers. The most essential of such devices are shown diagrammatically in Fig. 6-6. The functions to be served by these auxiliary devices can be listed as follows: (1) Circulation of saline water into and through the plant, together with the removal of the distillate and unevaporated brine.

(2) Removal of air and dissolved gases from the plant which enter with the saline water. (3) Maintenance of pressure and temperature distribution between effects. (4) Control of steam input to maintain a given water output.

The manner by which these functions are satisfied is illustrated in the figure. Circulation of the water into and out of the plant is accomplished by three pumps, namely, the seawater supply pump, the brine pump, and the distillate pump. These pumps are float-controlled to maintain given water levels in the first and last effects and in the final condenser hot well, respectively. The vacuum pump serves the dual purpose of removing air and dissolved gases from the system and maintaining a given pressure in the final condenser. Flow of the unevaporated brine from one effect to the next results from the pressure difference between the first effect and the last, with orifice restrictions between effects so proportioned as to give the desired pressure-temperature distribution. The steam input to the plant is regulated by an electrically controlled valve, actuated by an orifice or weir-type device placed in the distillate stream. Further regulation of the steam supply for any cutoff in the flow of water in the first effect, is maintained by a relay, controlled by a float or other water level indicator in that effect.

V. DESIGN CALCULATIONS

The design calculations for a multiple-effect distillation plant are based on the use of heat and mass balance equations for the determination of temperature and flow rates in the various parts of the system, and upon heat-transfer equations for the determination of heat-transfer areas. To illustrate the application of such equations for design calculations, the case of a three-effect seawater distilling plant, arranged as in Fig. 6-6, will be considered. In such a plant the direction of feed flow through the evaporators is forward. Seawater enters the plant as cooling water for the condenser, part of which is discarded after going through the condenser, and the remainder is circulated through the plant as distiller feed, heated on its way before entering the evaporators in the respective heat exchangers. The rate of production is assumed to be 1000 lb/hr of distilled water. Since the plant has three effects and a corresponding GOR of 2.4, its steam rate should be about 416 lb/hr. This steam must be divided between the vacuum pump and the first-effect heating element. A rule of thumb used for approximation purposes affirms that the vacuum pump in a distiller plant will require about 0.04 lb of steam for each pound of product. Thus, for an output of 1000 lb/hr of water it would mean that 40 lb/hr of steam must be reserved for the vacuum pump, leaving 376 lb/hr for the first-effect heating element. The temperature of the external steam to be supplied to the plant is set at 227 °F, and the cooling water input at 70 °F. The magnitudes of other temperatures in the system are determined as follows:

a. Maximum Brine Temperature. This temperature must be less than the condensation temperature of the external steam but should be as high as possible, and will depend upon the scale control method used, which in this example will be assumed to be an additive containing polyphosphates. Therefore, the maximum brine temperature, which is that in the first effect, will be limited to 190 °F.

b. Minimum Brine Temperature in the Evaporator. This value, which exists in the third effect, depends upon the vacuum to be maintained in the condenser and for distillers is likely to be 26 in. Hg, thus fixing the evaporation temperature of the brine in the third effect at approximately 125 °F.

c. Other Terminal Temperatures. The temperature of the outgoing cooling water from the condenser is assumed to be 85 °F, a value low enough to satisfy the local provisions for avoiding serious thermal pollution of the receiving water. Only part of this cooling water is discarded, the remainder being circulated through the plant as distiller feed. The temperature of the distiller feed and the cooling water are, therefore, the same, namely, 85 °F.

The temperature of the outgoing brine and the distillate are both assumed to be 95 °F, a 10 °F terminal temperature differential above that of the distiller feed, and from the viewpoint of heat exchanger area is necessary in order for heat transfer to occur between these outgoing streams and the distiller feed. The 95 °F temperature for brine is ecologically permissible because of its relatively small quantity as compared with the larger quantity of 85 °F cooling water with which it mixes.

Although the distillate temperature of 95 °F is too high for direct delivery to domestic service, it may be assumed that it will either be mixed with a large quantity of cool freshwater or be permitted to cool by atmospheric exposure in an open reservoir. When the distilled water is furnished directly to the domestic distribution system, the temperature must be lower than the 95 °F suggested above. In these cases the plant layout will differ from Fig. 6-6 in that it will provide for either a separate stream of cooling water to be used in the condensate cooler, or it places the condensate cooler ahead of the condenser so that the coolest water enters the condensate cooler. By these changes the temperature of the distillate can be brought within about 10 °F of the cooling water inlet temperature.

d. Temperature Changes in the Feedwater Prior to Entering the First Effect. It might seem desirable to have the feedwater enter the first effect at a temperature very close to 190 °F maximum temperature of the evaporating brine. However, since the final heating of the feedwater is accomplished by the condensation of vapor produced in the first effect, the final feedwater temperature must be less than the vapor temperature in order to absorb heat from condensation of that vapor. An inlet final feed temperature of 180 °F allows a satisfactory terminal temperature difference of 10 °F, and at the same time ensures that there will be no flashing of vapor in the water pipelines before the feedwater enters the first effect.

Heating of the feedwater from 85 ° to 180 °F is accomplished by the four heat exchangers shown in Fig. 6-6. The temperature rise in the brine and condensate coolers is given by the heat balance equation:

$$W_f c_{11}(t_{12} - t_{11}) = Wc_7(t_7 - t_8) + W_{bd} c_{18}(t_{18} - t_{19}) + \text{losses}$$

in which the subscripts refer to the state points on the figure. c represents the specific heat in the various streams. Because of the comparatively low temperatures in these heat exchangers the losses of heat therefrom will be small and will be disregarded in the equation. Scale control experience indicates that the salinity of the outgoing brine should not exceed twice that of the incoming seawater; hence W_{bd} will equal W. The brine temperature of 125 °F (t_{18}) leaving the third effect and entering the brine cooler will be at least four degrees higher than the condensate temperature, t_7, leaving the condenser, the difference being due to 2 °F BPE and 2 °F undercooling in the condenser to ensure complete condensation. Substitution of the above described temperatures in the foregoing equation gives the temperature of the feedwater leaving the two coolers as:

$$t_{12} = 112.9\,°F$$

The heating of the feed stream from 112.9 ° to 180 °F takes place in the two feed heaters shown in the diagram. It is assumed that each of the two heaters increases the temperature of the feedwater by an equal amount, i.e., by 33.55 °F in each.

e. Selection of the Brine Temperature in the Second Effect. This temperature is not too critical, inasmuch as slight changes in it will not alter the heat balance appreciably. It is usual to make it such that there are nearly equal temperature differences for heat transfer in the several evaporators, which is accomplished by having all of the evaporators of equal size. For a first trial in the above example, it can be assumed that the brine temperature in the second effect will be 158 °F, i. e., halfway between the brine temperature of the other two effects.

f. Calculation of First-Effect Output. The heat balance for this effect can be written as follows:

$$W_s(h_a - h_b) = W_f c_{14}(t_{15} - t_{14}) + \Delta W_1 (h_1 - h_{15}) + \text{losses}$$

where h represents the enthalpy and ΔW_1, the amount of vapor formed in the first effect. Values of enthalpy and specific heat are obtained from steam tables and charts, and when substituted in the equation give the value for the amount of vapor formed in the first effect as:

$$\Delta W_1 = 346 \text{ lb/hr}$$

g. Calculation of the Vapor Needed by the First-Effect Heater. This quantity is determined by a heat balance on the first-effect feed-water heater, which can be written as follows:

$$\Delta W_1'(h_1 - h_2) = W_f c_{13}(t_{14} - t_{13})$$

Again substituting numerical values for all quantities except $\Delta W_1'$, the value for the amount of vapor needed by the first-effect heater is found to be:

$$\Delta W_1' = 62.9 \text{ lb/hr}$$

h. Calculation of the Vapor Produced in the Second Effect. The heat balance for this effect can be written as follows:

$$(\Delta W_1 - \Delta W_1')(h_1 - h_2) + (W_f - \Delta W_1)c_{15}(t_{15} - t_{16}) = \Delta W_2(h_3 - h_{16}) + \text{losses}$$

where ΔW_2 is the vapor produced in the second effect. Again substituting numerical values from tables, charts, and previous calculations gives:

$$W_2 = 326.5 \text{ lb/hr}$$

TABLE 6-2

Tabulation of Calculations for the Three-Effect Distiller

Effect no.	Temperatures (°F)			Brine[a] salinity at outlet (%)	Boiling point elevation	Production rate (lb/hr)	
	Brine	Condensing vapor	Difference			GOR = 2.4	GOR = 2.37
1	190	227	37	4.24	1.1	346	354
2	158	188.9	30.9	5.28	1.0	326	334
3	125	156.5	31.5	7.00	1.5	308	316
					Total	980	1004

[a]Brine salinity at inlet to the first effect is 3.5.

i. Calculations for the Remaining Heat Balances. Heat balance equations were written for the third effect, the condenser, and the second effect feedwater heater, with terms similar to those in the previous equations. The results of these calculations are shown in Table 6-2. The total production rate is seen to be less than the 1000 lb/hr for which the design is intended. In order to obtain the desired output, the steam condensed in the first effect must be increased, thus increasing the vapor produced in each effect. By a series of modifications of the calculations, the figures in the last column, labeled "GOR-2.37," were obtained, based upon a steam rate of 421 lb/hr rather than the 416 lb/hr originally used.

At this point it seems relevant to present justification for the apparently arbitrary 0.8 factor used in relating the GOR to the number of effects. Although it is possible to express this relationship in algebraic terms for any specific arrangement of feedwater heaters and condensate and brine coolers, the algebra becomes rather involved and is avoided by using the simple 0.8 factor. Detailed algebraic analyses for single-, double-, and triple-effect plants designed for marine usage have been made by Ireland [5]. While the plants analyzed by him are not completely similar to the plant of Fig. 6-6 because of their application to marine service, nevertheless, the resulting equations will give close approximations for the GOR in the plant of Fig. 6-6. For instance, the marine plants considered did not include brine coolers, possibly reflecting the fact that the movement of the ship eliminated the temperature pollution aspect of such discharge. In place of this, two additional heaters appeared in the feedwater line, one being a heater placed in the vapor line between the last effect and the condenser and the other being the air ejector condenser, neither of which is included in Fig. 6-6. One further difference should be noted, namely, that the condensate cooler is placed in the seawater line ahead of the condenser, thus serving to cool the distillate to the lowest possible temperature. This is particularly important from a practical standpoint, since water for domestic consumption on board ship cannot have the cooling effects of dilution or atmospheric cooling available to land-based plants.

Ireland's derivations resulted in equations which have the following general form:

$$\frac{W}{W_s} = n\left(1 - \frac{(W_f/W)[n(t_5 - t_{11}) - \delta t] - 0.6(t_a - t_5)}{h_{fg,5} + (W_f/W)[n(t_5 - t_{11}) - \delta t]}\right)$$

where δt is the temperature rise in air ejector condenser and W_s is the steam supplied to first effect; subscripts refer to points on Fig. 6-6.

Under the conditions of the illustrative example, namely, three effects, steam supplied at 227 °F, the third effect operating at 125 °F and $W_f/W = 2$, the above equation gives a value of $W/W_s = 2.58$. If ejector steam is added at the rate of 0.04 lb per pound of distillate product, the value of the GOR is 2.37, which is slightly less than the 2.4 obtained by using the 0.8 factor. Thus is exemplified the fact that the use of the 0.8 factor can give nearly the same GOR as the more cumbersome equations presented above.

j. Calculation of Heat-Transfer Areas. The flow of heat in the several evaporators can be calculated using the following equation:

$$Q = UA\theta_m$$

in which Q is the amount of heat given up in the condensing coil, θ_m is the mean difference in temperature in °F, U is the overall transfer factor — Btu/(hr × °F × ft²), and A is the area of the heat-transfer surface in ft². The quantities of heat to be transferred are known from the previous calculations, together with the mean temperature differences. However, values for the overall heat-transfer factors must be determined for the particular configurations of tubing chosen for the plant. The designer generally has access to proprietary information on this point and can proceed to determine actual areas for the configuration used by a specific company. Lacking the specifications of the particular arrangement of heating surface, the product, UA, having the dimensions Btu/(hr °F), is indicative of the area variation. Table 6-3 gives this quantity for the several effects

TABLE 6-3

Heat Flow Data

	Q Btu/hr	θ_m °F	UA
Evaporator I	367,280	37.0	9900
Evaporator II	286,700	30.9	9250
Evaporator III	270,000	32.0	8440
Condenser	343,200	45.6	7520
Heater I	62,000	21.4	2900
Heater II	63,800	22.8	2800
Condensate cooler	26,000	9.02	2885
Brine cooler	27,600	10.5	2630

and other heat exchangers. Values for U, the overall heat-transfer factor, vary from about 100 Btu/(hr ft^2 °F) for the liquid-to-liquid units used for cooling the brine and condensate to 500 or more in the evaporators having smooth tubes under favorable conditions. The value of U may be as high as 3000 to 4000 for evaporators with corrugated tubes or with other means of enhancing the transfer of heat.

An examination of this table shows that the product, UA, decreases in successive evaporators. If the overall transfer factor was the same in the several effects, this would mean successively smaller heat-transfer surfaces in the second and third effects. However, it is generally found that the heat-transfer factors are less in each successive effect. Since this corresponds to the need for increased heat-transfer surfaces, it is usual to design a plant with evaporators of equal heat-transfer areas, balancing thereby the decreasing heat-transfer coefficients with the decreasing values of UA. It is also observed that the condenser surface requirement will be nearly the same as for the several effects. The two feedwater heaters will have areas only about one-third those of the evaporators, while the liquid-to-liquid units needed for cooling the distillate and brine may have area requirements nearly as much as for the evaporators, due to the lower heat transfer coefficients in those lower temperature devices.

The foregoing comments on design are intended to give some impression of the procedure for determining the major quantities during a preliminary design study. It should be clear from previous chapters that the design study normally would involve consideration of several alternative designs. In the case of multiple-effect plants, this would take the form of calculating the cost of water produced by plants having different numbers of effects. This detailed consideration would likely be handled using digital computers, since this makes possible the variation of many more of the variables than would be feasible without the computer.

REFERENCES

1. M.S. Sachs, Desalting Plants Inventory, Rept. No. 1, Office of Saline Water, U.S. Dept. Interior, Jan. 1, 1968, 25 pp.
2. D.D. Kays, Advances in Vertical Tube Plants for Sea Water Distillation, Advance Paper, Western Water and Power Symp., Apr., 1968, 27 pp.
3. L.A. Bromley, Multiple-Effect Rotating Evaporator, Desalination, 1 (No. 4), 1966, pp. 367-393.

4. J.T. Chambers, Performance Characteristics of Sea Water Distillation Equipment, Inst. of Eng. Res. Ser. No. 75, Issue No. 19, 1959, Univ. of Calif., Berkeley, California.
5. M.L. Ireland, Jr., Economic Features of Low-pressure Evaporating and Distilling Plants for Merchant Ships, Advance Paper No. 1, Annual Meeting Soc. of Naval Architects and Marine Engineers, Nov. 1946, 14 pp.

Chapter 7

MULTIPLE-STAGE FLASH DISTILLATION PLANTS

Since the early 1960s, all of the large plants constructed for producing freshwater from seawater have, to a greater or lesser degree, included in their design the multiple-stage flash distillation process. The general principles involved and the arrangement of the stage components have been described in Chapter 4. The purpose of the present chapter is to apply quantitatively these general principles to the design of actual plants.

I. FLASH DISTILLATION DEVELOPMENT

The modern version of multiple-stage flash distillation is comparatively new, having been invented in the middle 1950s, although before that time the principle of flash distillation had been known. The earliest application of this principle for the production of freshwater used arrangements resembling submerged-tube multiple-effect plants. Vapor, produced by "flashing" in one of the chambers, was piped to a second chamber where it was condensed on the outside of tubes through which cooling water flowed. Thus for each stage there were two separate chambers — a flash chamber for producing vapor and a condensing chamber where the vapor was liquefied. An example

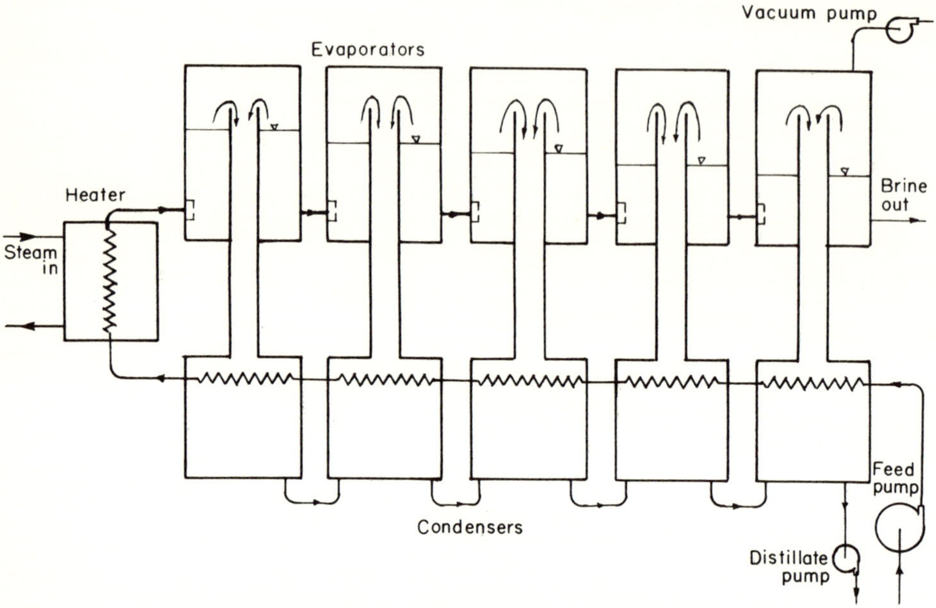

FIG. 7-1. Five-stage flash distiller with external piping. (Schematic only.) Note that each evaporator and condenser is in a separate vessel. Water pretreatment and vacuum systems are not shown.

of this early type of flash installation, shown diagrammatically in Fig. 7-1, is the NATO plant installed at Gibraltar in 1960. The five stages of this plant required five flash chambers, five condensers, and a brine heater, eleven separate components, all of which were enclosed in separate casings and interconnected to the others by piping external to the cases.

A refinement of this same type of design is seen in the 1957 Kuwait installation, in which most of the external piping is eliminated and several stages are enclosed in a single casing. This latter flash distillation plant, the production capacity of which is 2.5 mgd, consists of four identical units, each having four stages. The vertical towers of the four individual units are single casings which contain the four stages as layers, one above the other, with the high temperature stages at the top. The brine heater is located at the top of the tower and, through a pressure-controlling orifice, delivers the heated brine to the first stage. Vapor is produced here by expansion due to the drop in pressure at the entrance to the stage, and passes through a mesh separator into the condenser compartment where it comes into contact with the condenser tubes, is liquefied, and collected as distillate by draining by gravity into the condensing chamber of the next stage. Unevaporated brine from the first stage flows through a second orifice to the second stage where more vapor is produced, this process being repeated for each of the remaining stages. The seawater feed is introduced at the base of the tower and flows first through the condenser tubes of the fourth stage, and then in succession through the condenser tubes of the other stages in series, finally entering the tubes of the brine heater to receive the final temperature increment. In spite of the small number of stages and the relatively low GOR of its design, this plant successfully demonstrated many of the features that are characteristic of modern multiple-stage flash distillation plants.

A. Prevailing Developments

The aforementioned plants differ from more recent plants in that the latter contain far greater numbers of stages. The first plant in the United States using modern design was installed in 1960 at the Mandalay Power Station of the Southern California Edison Power Company for experimental purposes. This plant is shown in Fig. 7-2. It was designed to produce 100,000 gpd and had 26 stages arranged in two horizontal cylindrical cases. The 22 heat recovery stages were

FIG. 7-2. Photograph of 100,000 gpd plant with 26 flash stages, Oxnard, California. Note inverted siphon traps to conduct distillate from stage to stage without vapor loss.

TABLE 7-1

Representative Multiple-Stage Flash Distillation Plants[a]

Location	Capacity (mgd)	Number of units	Date installed	Number of stages	GOR[b]	Maximum brine temp. (°F)
Kuwait	2.4	4	1957	4	3.1	194
Oxnard, California	0.10	1	1959	26	6.0	195
Kuwait	2.4	2	1960	19	5.7	194
Isle of Guernsey	0.600	1	1960	40	10.6	204
Point Loma, California	1.0	1	1962	36	10.5	200
St. Thomas, Virgin Islands	0.275	1	1962	28	6.8	195
Guantanamo, Cuba	1.50	2	1964	15	6.6	195
St. Thomas, Virgin Islands	1.0	1	1965	26	8.0	250
St. Croix, Virgin Islands	1.5	1	1965	15	6.0	250
Malta	1.23	1	1967	24	8.7	192
Chula Vista, California	1.0	1	1967	68	20.0	250
Key West, Florida	1.0	1	1967	50	13.3	250
Terneuzen, Netherlands	7.66	2	1968	38	11.0	248
Rosarito Beach, Baja California	7.5	2	1969	44	10.6	235
Nassau, Bahama	2.4	2	1970		14.5	
Curacao, Netherlands Antilles	1.058	1	1970	38	10.0	
Kuwait	4.808	1	1970	26	8.0	195
Jedda, Saudi Arabia	5.0	2	1970	42	10.1	
Jersey Island	1.8	1	1970	47	12.5	
Kuwait	30.0	5	1971	25	8.0	194

[a]Numerical data in this table were obtained from Refs. [1] and [2].

[b]The GOR is the ratio of the weight of water produced to the weight of steam used.

contained in one case and the four heat rejection stages in the other. The housing of the heat rejection stages in a separate case was necessary in order to provide a means to remove at the exit of the case a part of the cooling water which would have to be discharged to maintain constant temperatures in the plant. This case, containing the four heat rejection stages, was divided into four equal sections by transverse partitions. The condenser tubes extended the full length of the case, passing through tightfitting holes in the transverse partitions. Trays beneath the condenser tubes were arranged to collect the condensate as it dripped off the tubes. This condensate was conducted from stage to stage through external loop seals. Brine flowed from stage to stage through orifices in the lower parts of the partitions.

The cylindrical case, containing the 22 heat recovery stages, was similarly arranged with two sets of condenser tubes running within its full length on each side of the casing which was divided into two equal halves by a vertical longitudinal partition. Ten vertical transverse partitions divided the case into eleven stages on each side of the longitudinal partition. The brine flowed through the eleven stages on one side of the longitudinal partition, made a U turn and returned, completing the 22 stages on the opposite side of the partition. The two sets of condenser tubes and condensate trays likewise conducted their respective flows through the 22 stages in series. For more than six years this plant was experimentally operated under various conditions of temperature and flow. At the end of this period it was moved to Catalina Island, where it is now operated occasionally to supplement the supply obtained from rainfall catchment.

Although the recent trend in multiple-stage flash distillation plants is for greater and greater numbers of stages, plants constructed since 1960 have the general basic features described above for the relatively small 26-stage plant. Table 7-1 lists a few of the significant multiple-stage flash installations. It lists only those supplying domestic and municipal potable water and does not include the many distillation units constructed as parts of steam power plants or industrial complexes such as for oil refineries, because of the basic difference in the criteria used in their design. The list in Table 7-1 is not intended to be a complete catalog of installation units, but rather to show their scattered location and their wide range of sizes. More complete lists of plants have been compiled by El-Ramley et al. [1] and by Sachs [2]. In El-Ramley's report he points out that plants of the multiple-stage type, including those serving power plants, industrial complexes, and potable consumption, have been increasing at a rate of nearly 30% per year from 1960 to 1970. This increase is more rapid than that of any other desalination process during the same period.

TABLE 7-2

Configurations Provided in 2.5 mgd Universal Plant Design[a]

Plant no.	Number of stages	Performance ratio lb water/1000 Btu	Heat transfer surface in ft^2	Steam required (lb/hr)
1	50	13.87	247,629	66,720
2	50	13.10	227,615	70,620
3	48	11.64	194,324	79,500
4	42	11.14	188,629	83,080
5	34	9.46	159,116	97,800
6	30	7.78	128,235	118,880
7	24	6.83	115,000	135,460
8	32	7.69	240,000	121,300
9	24	5.57	170,000	167,400
10	32	6.76	200,000	137,900
11	24	5.35	160,000	174,400

[a]Plants nos. 1 to 7 are designed for acid scale control, brine temperature 250°F. Plants nos. 8 to 11 are designed for polyphosphate scale control, brine temperature 190°F.

Data in table taken from OSW Saline Water Conversion Report for 1969-70.

According to publications by the U.S. Office of Saline Water, and also listed in Table 7-1, the distillation plant at Jedda, Saudi Arabia, chose the Universal Design for a 2.5 mgd plant. This design contains considerable flexibility in that eleven configurations are inherent in it, the variations being listed in Table 7-2. Selection of a particular plant configuration depends upon the local cost of steam, the fixed charge rate, and the type of feedwater pretreatment to be used. The very complete set of five manuals listed in the reference [3] not only makes possible a knowledgable selection of a Universal Design plant that would be appropriate for particular local conditions, but also makes available practical design information.

B. Unconventional Proposals

The plants listed in Table 7-1 differ in design details but are all quite similar in concept, i.e., the flashing brine flows horizontally through a series of orifices in stage partitions, and the vapor produced condenses on tubes through which the incoming brine is flowing and being heated. The cost of heat-transfer tubing in such multiple-stage flash distillation plants has been estimated to represent from 30 to 40% of the total cost of plant materials. Corrosion of this tubing due to seawater feed is likely to be serious enough to require replacement well before the failure of other plant components. These factors have led to several proposals for the use of unconventional multiple-stage flash distillers in which part or all of the heat-transfer tubing has been eliminated.

Othmer [4] proposed that the heated flashing brine be expanded in a vertical evaporator tower structure containing several stages placed one above the other. The vertical placement, with the high pressure stage uppermost, was so designed that gravity provided the stage pressure differentials. Brine feed water, issuing from the orifice at the top entry to each stage, would fall in a spray pattern, thereby flashing and providing ample surface area for disengagement of the vapor in each stage. In a parallel tower, cooled freshwater would be sprayed from the top into complementary stage compartments which are connected to each stage evaporator spray chamber by horizontal vapor passages. Since the cooling water in the second tower must pass upward through the stages countercurrently to the expanding brine in the first tower, it is necessary to provide a separate cooling water circulating pump for each stage. It is recognized that this design is basically different from the conventional design in that it eliminates the need for expensive and critical tubing. Although considerable research and experimental work has been spent on this scheme, no production plants of this kind have as yet been installed.

The above unconventional proposal eliminated the use of the costly and potentially corrosion-prone tubing which is the essential feature of conventional multiple-stage flash distillation plants. A further unconventional suggestion proposes to use tubing in the condenser section, but eliminates the vacuum equipment from the plant by using a humidification-dehumidification scheme. By this means the change in vapor temperature from stage to stage is accomplished through the automatic change in partial pressure of the vapor, rather than having the water pass through pressure-controlling orifices. A small experimental unit of this type was built by Hodges [5] at Puerto Penasco, Mexico. Originally constructed to use solar-heated water, it is now operated with waste heat in the cooling water from diesel engines.

This plant consists of two vertical towers with horizontal cross-connections at several elevations, one for each stage. One tower is packed with Rashig rings and through this the heated seawater passes downward countercurrently to air which is flowing upward. Contact between the warm water and air results in warming and humidifying the air. Controlled by orifices in the cross-connection lines, this humidified air flows across to the second tower in which there are banks of condenser tubes through which cooling water is flowing, the water being pumped upward from bank to bank. The humid air, coming into contact with the cool outer surface of the tubes, is cooled and partially dehumidified as the water vapor condenses and drops to the condensate trough. A fan in the bottom cross-connection draws the cooled air out of the condenser column and returns it to the evaporator column for reuse. Thus the device has most of the features of a multiple-stage flash distiller, the chief difference being that the change in partial pressure of the humidity takes the place of the pressure-controlling orifices between stages in the conventional plant. To offset the poor heat-transfer factor due to the presence and circulation of air, the Puerto Penasco plant has finned or spiny tubing in the condenser. The power required for this circulation of air is considerable. Heat energy is further lost because the circulation of the uncondensed portion of air through the system conveys heat from the evaporator to the condenser without involving this heat energy in evaporation. All of these factors indicate heat losses and are more or less offset by the cost and operating disadvantages of the vacuum system in the conventional plants. The scheme is still in the experimental stage and as yet is not widely used.

C. Basic Design Arrangements

The conventional type of multiple-stage flash distillers have three basic geometric designs for the arrangement of the condenser tubes relative to the direction of brine flow. If the brine is pictured as flowing along a horizontal channel in which submerged orifices control passage of the brine from stage to stage, it is conceivable that the condenser tubes could be either parallel to the direction of brine flow or perpendicular to it, in which latter case the tubes could be either horizontal or vertical.

The long-tube configuration, in which the condenser tubes are parallel to the direction of the brine flow, seems to be preferred by American designers and can be exemplified by the Point Loma Demonstration Plant constructed at San Diego for the [6] U.S. Office of Saline Water. The 36 stages of this plant were contained in five separate casings. Four of these casings were 80 ft long and contained eight stages per casing. The fifth casing was approximately one-half as long and contained four stages. In each long casing the condenser tubes were 80 ft long, passing through the seven transverse partitions which divided the casing into eight stages. Essential for the efficient operation of the process was the care that had to be taken to prevent leakage of vapor from stage to stage through the annular spaces around the tubes in the partitions. Leakage was minimized by close precision drilling of the holes through which the tubes had to be drawn from stage to stage. Additional resistance to leakage between the higher pressure stages was achieved by inserting bored ferrules into the stage partitions, so designed that the length of path for leakage flow was several times the thickness of the partition itself. Obviously the technique used for pulling the tubes through the partition holes between the several stages was the limiting factor in securing small clearances.

The major disadvantage of the long-tube configuration is the large amount of ground area required, and the need, during tube replacement, of having a clear space at the end of the casing which is at least as long as the tubes. In order to reduce this space requirement where ground area is limited, short-tube schemes have been developed in which the tubes for each stage are separate and are arranged perpendicular to the direction of brine flow. Such tube bundles can be either horizontal or vertical, the horizontal arrangement having the advantage that several brine channels can be placed one above the other, thus making for a very compact installation.

An example of the horizontal arrangement with stacked brine channels is the Malta plant [7]. This 24-stage plant was built with eight stages on each of three levels, one above the other. The 29-ft long condenser tubes of each stage were U shaped, and the tube bundles for all stages were arranged so that they could be removed from the same side of the casing for repair. Since these bundles were approximately 20 ft long, a clear space of only a little more than 20 ft was required alongside of the distiller. With this design it was possible to place the distiller plant inside of one of the power plant buildings, the ground area required being about 50 by 75 ft. This may be compared with the Point Loma layout which had approximately the same plant capacity but required a space of 80 by 160 ft.

As indicated earlier in this section, tube bundles perpendicular to the direction of brine flow can also be placed vertically. This arrangement has been used in several plants constructed in the United Kingdom, and also has the advantage of requiring less ground area than the long-tube arrangement described above. However, it does require sufficient height for the removal of the condenser tube bundles from the top of the evaporator. Its chief disadvantage lies in the problem of heat transfer, since it is more difficult to maintain a thin condensate film on a vertical surface than on the outside of a horizontal tube. This latter factor probably accounts for the smaller number of installations using this configuration.

In comparing these design concepts it should be noted that the long-tube arrangement has fewer tube ends that must be sealed to tube sheets. This results in less hydraulic resistance to flow and hence a saving in pumping power. In the long-tube design the stage length is controlled by the length of the condenser tubes, rather than by vapor release from the brine. In the cross-tube arrangement the stage length is determined by the vapor release from the brine, and must be long enough for the vapor formed to escape from the surface of the flowing brine. Comparing the Point Loma and Malta plants in this respect, it is noted that the former had a stage length of 10 ft whereas, in the latter, the stage length was only 5 ft. Thus, the advantage of the cross-tube design is that of requiring less space than the long-tube design. However, where space is not the critical factor, the long-tube design would have the advantage of less pressure loss due to friction in the tubes and water boxes, and simpler construction due to fewer tube end joints than in the cross-tube design.

II. THERMODYNAMIC ANALYSIS

The prediction of the performance of a multiple-stage flash distillation plant can be made by using thermodynamic relationships expressed in heat balance and mass balance equations. The sizes of the evaporator-condenser units are also estimated by using heat-transfer relationships. The formulation of the equations connecting these various quantities and relationships will be presented in two parts, the first being based on average values of properties and the second involving point-to-point changes in fluid properties as they would be used in preliminary design.

Basic temperature and flow rate conditions in a simple flash plant are shown in Fig. 7-3. The upper diagram shows the temperature dis-

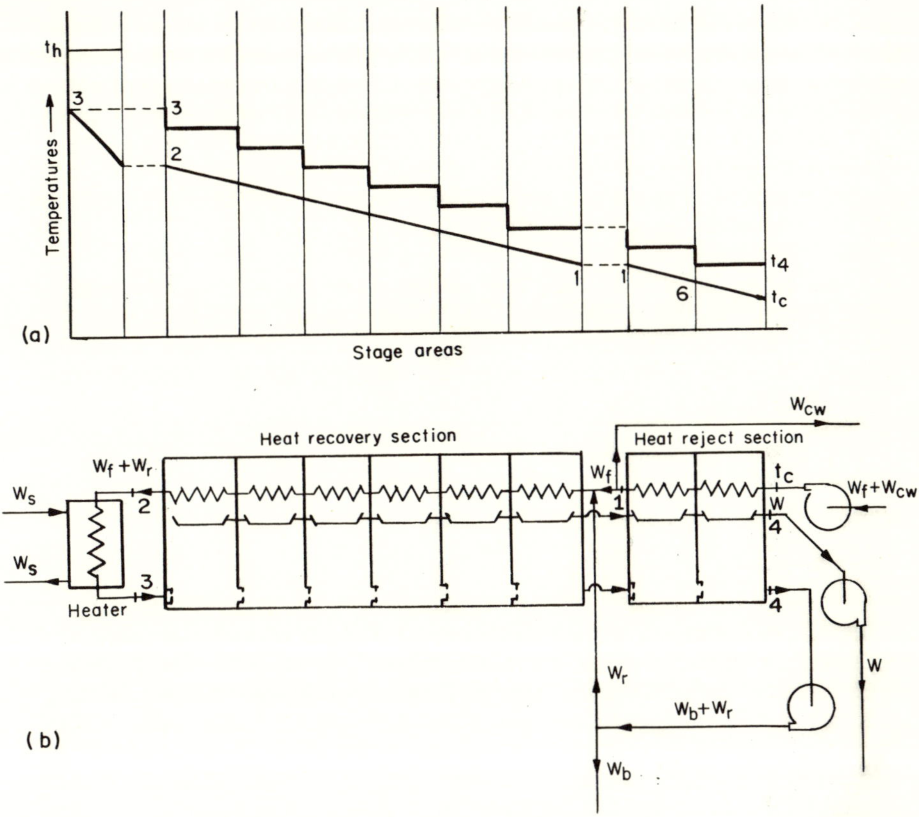

FIG. 7-3. Modern multiple-stage flash distiller diagrams. (Schematic only.) (a) Temperature diagram. Numbers on this diagram correspond to state points used in the analysis. (b) Flow sheet. Flow rates shown correspond to those used in analysis.

tribution, and the lower diagram shows the several flow rates. The highest temperature in the plant is t_h, the temperature of the condensing steam in the brine heater. Seawater enters at the lowest temperature, t_c, and is heated up to temperature, t_1, by condensation of product water in the several stages of the heat rejection section. At this point most of the seawater, or cooling water W_{cw}, is discharged from the plant for heat rejection and the balance, W_f, is mixed with some of the brine, W_r, flowing out of the last stage of the heat rejection section. This mixture of warm seawater and recycled brine, $W_f + W_r$, is pumped through the tubes of the heat recovery section where it is heated to temperature t_2. From there it is passed through the brine heater, where it reaches its maximum temperature, t_3, and then cascades downward through the several stages, flashing off some product vapor in each stage and leaving the last stage at temperature t_4. The elevation of temperatures t_1 and t_4 above the inlet seawater temperature, t_c, is regulated by civil law to levels which will not destroy the biological and ecological balance of the waters into which the brine and cooling water are discharged – generally an elevation of not more than 15° to 20°F.

The lower portion of Fig. 7-3 shows water flow rates. As indicated above, the total amount of distillate vapor, W, reduces the amount of the brine mixture, $W_f + W_r$, to the amount of the brine discharge, $W_r + W_{bd}$, the latter term representing the brine blowdown. This relationship is expressed by the equation:

$$W_f + W_r - W = W_r + W_{bd}$$

or

$$W_f = W + W_{bd}$$

Since W represents the predetermined distillate production rate, it is important that the brine discharge rate be such as to maintain the concentration of scale-forming ions brought in by the feedwater, W_f, at a value low enough to prevent scale formation. It has been found that this balance can be maintained when the brine blowdown rate is approximately equal to the distillate rate, $W = W_{bd}$. Therefore, from the above equation it follows that $W_f = 2W$.

The flow rate of cooling water, W_{cw}, is determined by the amount of heat to be rejected. Recalling that all of the heat energy added in the brine heater must be eliminated from the plant, it is seen that W_{cw} will be determined by the elevation of temperature t_1 above the ambient water temperature, t_c, as well as by the amount of heat discharged with the waste brine and product. For conservative design it is usual to ignore the heat dissipated from the surface of the equipment to the air about it, because this latter quantity is generally only 1 or 2% of the heat input. Thus the significant rate of cooling water will be determined by an equation representing the overall heat balance in the plant:

$$W_{cw}c(t_1 - t_c) = Q - W_{bd}c(t_4 - t_c) - Wc(t_4 - t_c)$$

where Q is the rate of heat input and c is the specific heat.

A. Analysis Based on Average Values

1. Heat Balance Equations

The following analysis and equations, patterned after the work of Chambers [8], assume that each stage contains the same heat transfer area, ΔA, and produces an equal amount of distillate, ΔW. By using average values for specific heat, c, and latent heat, h_{fg}, it is possible to write equations which describe the heat balance for different portions of a distillation plant. The values for specific heat, c, in the following equations are average specific heats for the range of temperatures and concentrations involved in the process. Their variation is small, slightly less than 0.07, being 0.93 for c_6 and 1.00 for c_5. The numerical subscripts designate points on the figure. The letter subscripts refer to the following: f, feedwater; cw, cooling water; bd, brine blowdown; r, recirculated brine; s, steam.

Heat balance on the coolant stream

$$Wh_{fg} = (W_{cw} + W_f)\,c(t_1 - t_c) + (W_f + W_r)\,c(t_2 - t_1) \tag{1}$$

Heat balance on the flashing stream

$$Wh_{fg} = (W_f + W_r)(t_3 - t_4)c \tag{2}$$

Heat balance on the brine heater

$$W_s h_{fg} = (W_f + W_r)(t_3 - t_2)c \tag{3}$$

Heat balance on the entire plant

$$W_s h_{fg} = W_{cw} c(t_1 - t_c) + W_{bd} c(t_4 - t_c) + Wc(t_4 - t_c) \tag{4}$$

In actual practice within a plant, the rate value of $W_{cw} + W_f$ is often very nearly equal to that of $W_f + W_r$, so that the relationship expressed in Eq. (1) may be simplified as:

$$Wh_{fg} = (W_f + W_r)\,c(t_2 - t_c) \tag{1a}$$

Equations (1a) and (2), when combined, give the result:

$$t_2 - t_c = t_3 - t_4 \quad \text{or} \quad (t_4 - t_c) = (t_3 - t_2) \tag{5}$$

This equation indicates that the temperature rise through the brine heater is equal to the terminal temperature difference between the

TABLE 7-3

Effect of Recovery Ratio on Plant Economy for an Idealized MSF Distiller[a,b]

Recovery ratio $W/(W_f + W_r)$ (1)	Steam economy W/W_s (2)	Brine temperature t_4 in °F (3)	Minimum number of stages (4)	Area economy W/A[c] (5)	θ_m (6)
0.050	0.925	138.5	-	12.4	53.9
0.060	1.39	126.6	-	10.8	43.0
0.080	3.16	104.8	3.2	11.3	24.4
0.090	7.75	93.9	9.5	4.88	10.3
0.092	9.64	91.8	12.2	3.90	
0.093	11.0	90.6	14.5	3.37	
0.094	12.9	89.5	17.4	2.86	
0.095	15.0	88.5	21.2	2.39	
0.096	18.6	87.3	28.3	1.81	
0.098	31.4	85.2	66.7	0.79	
0.100	102.5	83.0	-	-	
0.101					
0.093[d]	9.1[d]	93.0[d]		3.70[d]	

[a]Computed from Eqs. (6) and (7).

[b]Conditions: $t_3 = 192\,°F$; $t_h = 218.5\,°F$; $t_c = 82\,°F$; $h_{fg(s)} = 966$; $h_{fg(c)} = 1014$; $c_p = 0.931$; $c_{p,c} = 0.942$; $U = 500$.

[c]Area economy figures are based on infinite number of stages.

[d]These comparative values are determined from published design data on the Malta plant [7].

waste brine and the incoming seawater. In a plant with an infinite number of stages, this would also be the constant temperature difference between the coolant stream and the flashing brine stream. For a finite number of stages, the average temperature difference between the two streams would be less than $t_3 - t_2$.

2. Steam Economy

As noted in previous chapters, steam economy can be expressed in terms of the gained output ratio, W/W_s, in which W_s is the rate of steam required for the distiller. However, this is not necessarily the total steam, W_s', required in a plant whenever steam is the energy motivating jet pumps, the steam of which does not return to the distiller. When such pumps are used, an adjustment for the steam needed for their operation, W_j, must be added to W_s in order to obtain a true GOR, $W/(1 + W_j/W_s)W_s$, in the final calculations. The accompanying equations refer only to the basic relationship, W/W_s.

The arguments used in obtaining the value of W/W_s is based on Eq. (2) from which it can be concluded that

$$t_4 = t_3 - \left(\frac{W}{W_f + W_r}\right)\left(\frac{h_{fg}}{c}\right) \tag{6}$$

Assuming that t_4 is always greater than t_c, and combining Eqs. (3), (5), and (6), the following equation is significant:

$$\frac{W}{W_s} = \frac{W}{W_f + W_r} \cdot \frac{h_{fg}}{c} \left[(\Delta t - \Delta t') - \frac{W}{W_f + W_r} \cdot \frac{h_{fg}}{c}\right]^{-1} \tag{7}$$

in which $\Delta t = t_h - t_c$ and $\Delta t' = t_h - t_3$. From Eq. (7) it is seen that the apparent gained output ratio, W/W_s, is a function of the recovery ratio, $W/(W_f + W_r)$, for given values of t_h, t_3, and t_c. It also indicates that there is a limiting value of $W/(W_f + W_r)$ for which the third factor on the right-hand side of the above equation becomes infinite. To illustrate this point, values of W/W_s have been computed from Eq. (7), using the temperatures set forth in the published design of the Malta plant [7], namely, $t_h = 218.5\,°F$, $t_3 = 192\,°F$, and $t_c = 82\,°F$, so that $\Delta t = 136.5\,°F$ and $\Delta t' = 26.5\,°F$. Values for the heat capacities and latent heats were obtained from charts and tables, as discussed in Chapter 2. The results of these calculations are given in Table 7-3, together with values of t_4

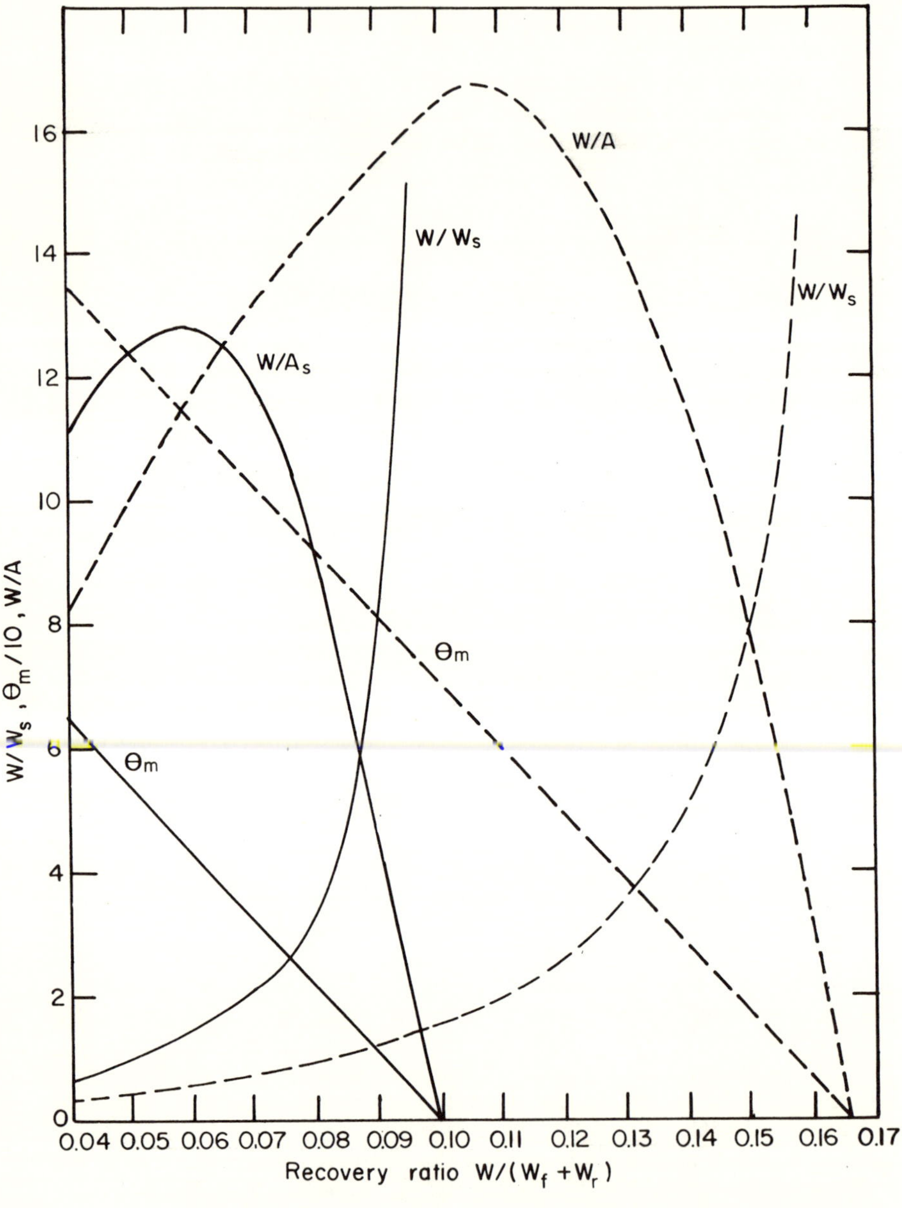

FIG. 7-4. Performance curves for MSF distiller with many stages. Conditions for solid curves: maximum brine temperature, 192°F; cooling water temperature, 82°F. Conditions for dashed curves: maximum brine temperature, 250°F; cooling water temperature, 70°F.

computed from Eq. (6). Also included are figures from the Malta plant design. Comparison of the Malta design figures with those computed from the theoretical equations shows reasonable agreement, the difference between the actual and theoretical steam economies being due in part to the inclusion of steam for the jet-type vacuum pump in the Malta figures as well as to the many idealizations inherent in the equations.

Values of W/W_s are plotted on Fig. 7-4, using the recovery ratio, $W/(W_f + W_r)$, as the horizontal coordinate. This figure shows that the curve for W/W_s is very nearly hyperbolic in shape, rising steeply and asymptotically toward the value of $[W/(W_f + W_r)]$ which makes the denominator zero in the third factor of Eq. (7). The physical significance of this denominator becoming zero is that the cooling effect of evaporation, equal to $h_{fg}W/(W_f + W_r)$, has reduced the temperature of the flowing brine from t_3 to t_c, the temperature of the incoming seawater. However, since the heat balance on the entire plant, Eq. (4), states that t_4 must be greater than t_c, the brine can never be cooled to the latter temperature, thus accounting for the asymptotic shape of the curve. Two facts of practical importance can be deduced from this curve. The first is that large values of steam economy, W/W_s, are to be found in a very narrow range of recovery ratios from about 0.09 to 0.095. In other words, this reflects the fact that in a plant of the above specifications the total circulating flow of brine can vary by only the small difference 10.5 to 11.1 lb of brine per pound of distillate. The second fact is that the steepness of the curve in this area is such that the steam economy of a plant is very sensitive to changes in brine flow rates; in other words, small changes in the brine flow rate will change the steam economy by a large amount. While this deduction is based on the idealized case of a plant with an infinite number of stages, these facts are borne out in practical plants.

3. Determination of Area Economy

Values for the area economy column, W/A, of Table 7-3 are based on arguments similar to those noted above. W/A is a function of the same variables as for steam economy, i.e., $W/(W_f + W_r)$, t_c, and t_3, as well as a function of the number of stages, n, the mean temperature difference for heat transfer, θ_m, the overall heat transfer coefficient, U, the boiling point elevation, α, and the area per stage, ΔA. Assuming that the areas of heat-transfer surfaces and temperature differences are equal in the several stages, the following equation for heat-transfer rates can be written:

$$Wh_{fg} = U(n\Delta A)\theta_m \tag{8}$$

Since temperature differences for heat transfer in the condenser section of the evaporator will be small, an arithmetic average can be used rather than the logarithmic mean. This temperature difference for the nth or last stage can be written:

$$\theta_m = [(t_4 - t_c - \alpha) + (t_4 - t_6 - \alpha)]/2 \tag{9}$$

Noting that

$$t_6 = t_c + (1/n)(t_2 - t_c) \tag{10}$$

and solving Eqs. (1a) and (2) for t_2 and t_4, gives the result:

$$\theta_m = (t_3 - t_c - \alpha) - \left(\frac{2n+1}{2n}\right)\left(\frac{W}{W_f + W_r}\right)\left(\frac{h_{fg}}{c}\right) \tag{11}$$

The heat-transfer rate in the brine heater can be written as:

$$W_s h_{fg} = UA'\theta_m' \tag{12}$$

in which the superscript ' refers to the conditions of the brine heater. The mean temperature difference in the brine heater, θ_m', will be larger than that in the stages, and so will be expressed as the logarithmic mean:

$$\theta_m' = \frac{t_3 - t_2}{\ln [(t_h - t_2)/(t_h - t_3)]} \tag{13}$$

Using the value of t_2 from Eq. (1a) gives the result:

$$\theta_m' = \frac{(t_3 - t_c) - [W/(W_f + W_r)](h_{fg}/c)}{\ln\left(\dfrac{t_h - t_c - [W/(W_f + W_r)](h_{fg}/c)}{t_h - t_3}\right)} \tag{13a}$$

Dividing Eq. (12) by Eq. (8), and using the value of θ_m' from Eq. (13a), gives the following for the area of the brine heater:

$$A' = \frac{n\Delta A\theta_m}{[W/(W_f + W_r)](h_{fg}/c)} \ln\left(\frac{t_h - t_c - [W/(W_f + W_r)](h_{fg}/c)}{t_h - t_3}\right) \tag{14}$$

The total area, $A' + n\Delta A$, needed for heat transfer in the plant then is:

$$A = n\Delta A\left[1 + \frac{\theta_m}{[W/(W_f + W_r)](h_{fg}/c)} \ln\left(\frac{t_h - t_c - [W/(W_f + W_r)](h_{fg}/c)}{t_h - t_3}\right)\right] \tag{15}$$

Combining Eq. (15) with Eq. (8) gives the area economy:

$$\frac{W}{A} = \left(\frac{U}{c}\right)\left(\frac{W}{W_f + W_r}\right)\theta_m$$

$$\times\left[\frac{W}{W_f + W_r}\,\frac{h_{fg}}{c} + \theta_m \ln\left(\frac{t_h - t_c - [W/(W_f + W_r)](h_{fg}/c)}{t_h - t_3}\right)\right]^{-1} \tag{16}$$

Since θ_m in Eq. (11) is a function of the number of stages, n, it follows that the area economy, W/A, is likewise dependent upon the number of stages. When the number of stages is very large and approaches infinity, the quantity $(2n + 1)/2n$ approaches unity, and θ_m is then a function of the recovery ratio only. Values of θ_m have been computed for $(2n + 1)/2n = 1.0$ and the temperature conditions assumed in Table 7-3. The values of θ_m were then substituted in Eq. (16) to obtain values for W/A, the results being given in column 5 of Table 7-3. These values are plotted on Fig. 7-4, showing their complete range of magnitudes, under the highly idealized assumption of an infinite number of stages.

Inspection of the curves for θ_m and W/A gives verification of conclusions which could be drawn from the equations. Equation (11) indicates that the relationship of θ_m to the recovery ratio is linear, with θ_m decreasing as recovery ratio increases and becoming zero when the cooling effect of evaporation is large enough to decrease t_4 to t_c, a situation which is not quite attainable due to the need for heat rejection. Deductions from Eq. (16) are less obvious, due to the rather involved nature of the denominator of the last factor. However, when the several factors on the right-hand side of Eq. (16) were evaluated in the course of computing W/A, it was found that variations in the denominator were small compared to the variations of θ_m and $W/(W_f + W_r)$. If the variations of this denominator are ignored, and the linear formula for θ_m substituted in Eq. (16), the result agrees with the parabolic shape observed on the curve of Fig. 7-4. It is worth noting that for the temperature conditions basic to Table 7-3, the maximum value of W/A on the curve corresponds to a value of about 2 for W/W_s, which indicates a steam economy far less than any modern plant would show. Also the table shows that for the same conditions, the temperature of the cooling water, t_4, would be excessively high, 120 °F. Thus, the achievement of high area economy is not accompanied by high steam economy or practical cooling operation. Increases in area economy for a given steam economy can come only from improved heat transfer.

It could be argued that the above conclusions based on the temperature conditions of the rather conservative design of the Malta plant might not be typical of the larger and higher temperature plants operating now and proposed for the future. For that reason calculations have been made based on the temperature conditions of the Claire Engle experimental plant [9] in San Diego, California. Its use of steam at 265 °F, maximum brine temperature of 250 °F, and cooling water at 70 °F is representative of advanced modern design. Using the above temperatures in Eqs. (7), (11), and (16), values were computed for W/W_s, θ_m, and W/A for various values of $W/(W_f + W_r)$. The results are plotted by dashed lines on Fig. 7-4, so as to distinguish them from the curves for the Malta plant conditions. Inspection of the dashed curves shows them to be similar in shape to the previously discussed curves, but displaced to the right; i. e., toward higher values of the recovery ratio. A line for θ_m is nearly parallel to that for the Malta conditions because the slope of this curve, h_{fg}/c, can vary only slightly for the entire range of conditions conceivable for the cooling of the brine in this type of plant. The curve for W/W_s is asymptotic to the vertical line at recovery ratio = 16.6%, so that values of steam economy high enough to be attractive for practical design occur at recovery ratio values in the small range of 0.155 to 0.16. This is in agreement with the fact that the fraction of the water vaporized during expansion of the water from the maximum temperature to the minimum depends on the temperature range. Thus, increasing the temperature range increases the fraction of vapor formed and hence the amount of water recovered. A further result of this increase of maximum brine temperature is that the amount of brine recirculated must decrease if the concentration of the discharged brine is to remain at a fixed value. It is further noted that the maximum point of the W/A curve corresponds to a very inefficient steam economy, and occurs at a value of recovery ratio too low to have practical significance. The portion of this curve in the range of the practical recovery ratio, 0.155 to 0.16, is nearly parallel to the corresponding portion of the similar curve for the Malta plant conditions. From these comparisons it appears that the qualitative conclusions drawn with respect to the variation of area economy ratio and steam economy ratio can be considered valid for all multiple-stage flash distillation plants, although the numerical values will depend on the temperature conditions encountered in the particular plant.

4. Determination of the Number of Stages

Obviously, an evaporator with an infinite number of stages, as assumed in the previous section, is only a theoretical consideration. Equation (16) and Fig. 7-4 indicate maximum limits for area economy values, relative to recovery ratios. It is evident that these values will be larger than can be realized by a finite number of stages, using the same recovery ratios. The selection of the number of stages to be used in a given plant is somewhat arbitrary, but should be greater than a "critical" minimum number. This minimum number is reached when the flashing range in each stage is increased to a value such that the temperature drop for heat transfer between the condensing steam and the brine in the condenser tubes approaches zero at the point where the brine passes to the next stage. In terms of the diagram of Fig. 7-3, the minimum number of stages is based on the expansion of the brine in the first stage from t_3 to t_2 plus the boiling point elevation, i.e., from t_3 to $(t_2 + \alpha)$. The resulting temperature drop, $t_3 - (t_2 + \alpha)$, will be the same for the other stages, so that dividing this value into the total temperature drop available, $(t_3 - t_4)$, will give the smallest number of stages which could, theoretically, heat the incoming water from t_1 to t_2 for given conditions of cooling water inlet temperature. Values for minimum number of stages were determined by this means, based on the conditions of Table 7-3 and the results shown in column 4. These values for minimum numbers of stages are smaller than would be practical, since the temperature differential between condensing steam and cooling water is zero at one end of the condensing tubing. This would thereby demand extremely long condenser tubes, the area of which would be excessively costly.

For numbers of stages exceeding the above theoretical minimum number, Fig. 7-5 indicates a diminishing gain in area economy as the number of stages is increased. The curves of this figure are based on conditions of Table 7-3. Using recovery ratios of 0.093, 0.090, and 0.088, three sets of values for θ_m were computed from Eq. (11). These values of θ_m were then used in Eq. (16) to determine the magnitude of area economy. The three resulting curves are similar in shape, indicating that the area economy increases rapidly in the range of low numbers of stages, but approaches asymptotically the theoretical minimum value for an infinite number of stages. From this set of curves it will be clear that there is a diminishing gain in area economy to be realized as the number of stages is increased. For example, using the lower curve, the gain in area economy to be realized by increasing from 15 to 20 stages is nearly 70% greater than that resulting from an increase from 20 to 25 stages.

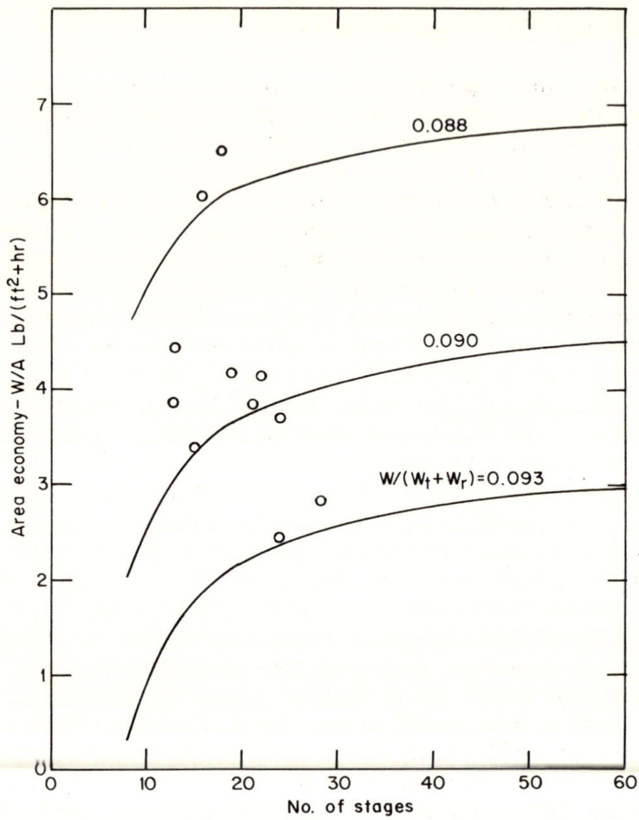

FIG. 7-5. Area economy of MSF distillers. Conditions for curves: maximum brine temperature, 192 °F; cooling water temperature, 82 °F; overall heat transfer coefficient, 500 Btu/(°F × ft^2 × hr). Note that points shown represent operating plants.

In preparing a plant design it is important to optimize total cost, considering several component costs, particularly W/A. Values for W/A have been computed for a number of plants and these values are plotted on Fig. 7-5. It is noted that all of the W/A points for actual plants fall between the curves of the figure. This was to be expected because the temperature conditions of the plants were almost the same as those assumed for the curves and the steam economy ratios lay within the assumed conditions, i.e., 4.6 and 11.

Since the selection of the number of stages is based on considerations other than that of achieving maximum area economy, El Ramley et al. [1] made a statistical study of the number of stages used in 164 flash distillation plants constructed prior to 1970 in an attempt to establish design trends. Their studies showed that there was a definite statistical relation between the number of stages and the capacity of the plants studied, and that this relation could be represented by the linear equation:

$$n = 19 + 6\,\text{mgd} \tag{17}$$

Equation (17) can be useful for the preliminary selection of the number of stages in the beginning of a design study, but it should be recognized that this equation is merely the statistical average of the conclusions reached by a comparatively small number of designers for plants in the range of 0.1 to 4 mgd. When applied to the Malta plant design, the number of stages computed from Eq. (17) is 26.2, a number that is in fair agreement with the number 24 used by the designers of that plant.

5. Maximum Brine Temperature

In a multiple-stage flash distillation plant, the maximum brine temperature, t_3, determines the amount of distillate which can be obtained from flashing a given amount of brine, $(W_f + W_r)$, to a given final temperature, t_4. This conclusion merely states in words the essence of Eq. (2) — that the higher the temperature, t_3, the greater will be the amount of distillate which can be produced. In the past, t_3 has been limited to about 250°F because of the difficulties of controlling scale formation. Since it is likely that scale control will be vastly improved in the future, it seems relevant to inquire whether the economy of multiple-stage flash distillation can be increased indefinitely by increasing temperature t_3, a question which has been studied by many investigators, and which obviously could greatly affect cost considerations.

The most general report concerning this question is one by MacLeod et al. [10]. In their study they seek to optimize the cost components of the multiple-stage flash distillation operation and they consider a wide range of variable conditions. They point out that for a given GOR, the basic cost of water distillate will depend primarily on (1) the per unit cost of steam necessary to raise the brine to the maximum temperature, t_3. However, to this cost must be added other

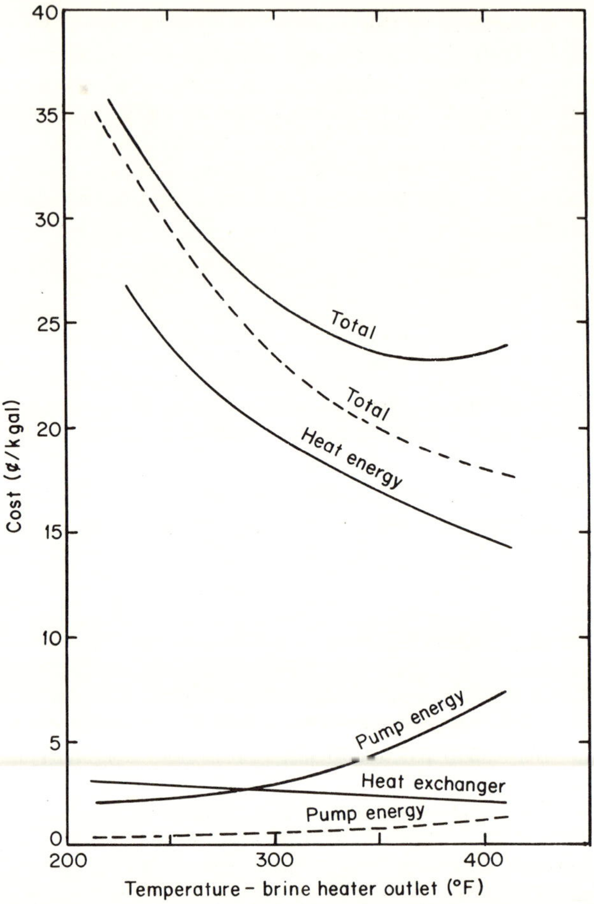

FIG. 7-6. Effect of maximum brine temperature on cost of water. Solid curves are for motor-driven pumps. Dashed curves are for turbine-driven pumps, with steam from turbine exhausting to brine heater. Cost of steam, \$0.55 per million Btu; cost of electricity, 0.007 per kWh.

significant considerations, namely (2) the area economy and amoritization cost of equipment per unit area of heat-transfer tubing, and (3) the amount and unit cost of electrical energy for pumping, etc., furnished from an external source. They seek to analyze this latter economy by considering the situation that would exist in two extreme examples. The first is one in which all pumping energy and electrical power is supplied from an external source, and a second one in which pumping energy is supplied by a back-pressure turbine which discharges all of its steam to the brine heater. It is assumed that the cost of labor and overhead will be essentially the same for both cases, so that in this analysis only the above three elements need to be considered.

For the case of the externally supplied electrical power, MacLeod's analysis is based on a unit cost of steam of \$0.55 per million Btu and a unit cost of electricity of \$0.007 per kWh. Their conclusions also assumed that the chief variable component of the plant cost would be the amount of heat-transfer tubing required, the unit capital cost of which was placed at \$2.32 per ft^2. The amortization rate was 7.4% per year. Using these figures and computing the above three costs of steam, electrical power, and heat-transfer tubing, they showed that the heat cost decreased as the maximum brine temperature was increased, that the electrical power for pumping increased very considerably, and that the heat-transfer tubing cost changed relatively little. The final conclusion from their computations was that the minimum value of these water cost components would occur at a maximum brine temperature somewhere between 350° and 400°F. The relative importance of the three elements considered in their conclusions is indicated by the curves in Fig. 7-6, in which plant costs are plotted against temperature.

The second example they considered is that in which the power for pumping, etc., is generated by a back-pressure turbine, operated by steam from the power plant and which exhausts its waste steam to the brine heater. In this example the heat and heat-transfer tubing costs will be the same as for the former case. However, in spite of the fact that the power generated in the turbine has the same common heat source as that used to heat the brine to temperature t_3, it is necessary to assign to the turbine-generator a cost equivalent to its capital and amortized cost, which is approximately 20% of the electrical cost of the purchased electricity of the previous example. This power cost is plotted against temperature by the dashed line in Fig. 7-6. The total cost of the three water components is also plotted in the figure. It can be seen in this chart that the total cost curve is shifted to the right as compared to the comparable curve of the previous example, suggesting the possibility of a maximum brine temperature of about 400°F. While it is clear that the exact value of brine temperature that could produce water at minimum cost will depend upon the values of many variables, this chart serves to indicate that the brine temperature could be advantageously increased to as much as 400°F, but that any increase beyond that level will not in itself improve the economy of water production.

B. Stage-by-Stage Analysis for Preliminary Design

The analysis using average values, as given in the previous section, is useful in discussing the general features of multiple-stage flash distillation plants, but contains a number of simplifying assumptions which must be reconsidered in undertaking the design of any specific plant. It was assumed that all stages could be represented by an average stage and that fluid properties could be evaluated at average values. These assumptions are not sufficiently accurate for design, so that calculations must be made on a stage-by-stage basis in which local variations of fluid properties and flow conditions are considered. Design analysis is normally done in two steps, with the preliminary design establishing the flow rates, temperatures, and areas required for heat transfer, and the final design setting up the physical arrangements needed to accommodate the flow rates and maintain the temperature distribution of the preliminary design. In this book only the steps in the preliminary design will be considered, since the final design is dependent upon the proprietary information available to the manufacturer.

Since design calculations are repetitive in nature and quite voluminous, it is usual for them to be made by means of digital computers. Computer programs for this purpose have been originated by a number of manufacturers for use by the company designers, and a few general programs have been published, examples of these being given in references [11-13]. While the calculations outlined in this chapter will be presented as though carried out manually, the reader should keep in mind the potential use of the equations involved in setting up a computer program for the same purpose.

The preliminary specifications for commencing the design of a distillation plant would generally include the capacity of the plant, the steam, and cooling water inlet temperatures and either the desired economy or the cost of water. Additional constraints may include the specifications of the maximum discharge temperatures of the product and brine streams and the salinity of the discharge brine. With numerical values for these quantities the designer is able to proceed with a preliminary design.

Assuming that the desired economy is given, the use of the analysis of the foregoing section will yield an approximate value for the recovery ratio, which, when combined with the maximum value of brine salinity at discharge, gives the rate of brine circulation and the rate of flow of the feed stream. The rate of flow of the cooling water can be determined from the GOR and the terminal temperatures of the waste brine and condensate, with the discharge temperature of the cooling water assumed equal to that of the waste brine. Thus, all of the flow rates are determined.

With flow rates and terminal brine temperature determined, and the number of stages selected, the production rates from the flashing brine can be calculated. This is done for each stage, beginning with the first or high-temperature stage, by writing a heat balance equation between the flashing brine and the vapor produced therefrom. The sum of the production rates of the several stages must equal the total desired production rate for the plant. If the first calculation does not meet this criterion, the brine flow rate must be adjusted and the calculations repeated until there is satisfactory agreement between the calculated and desired production rates. In making these calculations, needed values for properties can be obtained as outlined in Chapter 3.

The next step is the determination of the temperature change in the seawater and brine as it progresses through the condenser tubing from stage to stage. Beginning with the last stage, the temperature rise of the seawater in that stage is determined by a heat balance between the seawater in the tubes and the vapor condensing on the outside of the tubes. To the vapor coming from the flashing brine in this stage must be added the vapor flashing from the condensate as it cools from the condensing temperature of one stage to that of the next stage. This quantity will be small in the high-temperature stages of the plant but may be as high as 10% of the vapor formed from the brine in the last stages of the plant. With this augmented vapor quantity the temperature rise of the seawater for the last stage is determined. This process is repeated for each stage, giving finally the temperature of the brine entering the brine heater. With this temperature and the predetermined maximum brine temperature, the amount of external steam for the brine heater can be computed, thereby checking the GOR. Should this check be unsatisfactory, the calculations must be adjusted until satisfactory agreement is reached.

Having determined all the temperatures and flow rates, the designer is now ready to estimate the condenser tube area. Using the heat balances from the preceding paragraph, the quantity of heat to be transferred in each stage is calculated. The logarithmic mean temperature difference for each stage is computed from the vapor and brine temperatures. To determine the area of condenser tubes requires that the overall heat-transfer coefficient must be known, and values for this quantity are generally based on the proprietary information available to a given company.

After the condenser tube area is determined, the size, number, and length of tubes can be fixed. Condenser tubes are generally made from copper alloys or other corrosion-resistant metals, and are fabricated by drawing to render them seamless. Sizes are specified by the outside diameter in fractions of an inch, and by wall thickness in terms of a sheet metal gauge number. Tube diameters used in distillation plants have a range from 3/8 to 1-1/4 in., but diameters of 5/8, 3/4, and 1 in. are the most common. Wall thicknesses of tubing have ranged from 16 to 22 BWG, with the majority of the installations using 18 BWG, a tube wall thickness equivalent to 0.049 in.

Having selected the size of tube to be used, the number of tubes required to convey the brine is calculated from the flow rate of the brine and the desired degree of turbulence to ensure effective heat flow. The degree of turbulence is specified in terms of Reynolds' Number, or water velocity in the tubes. For the generally used tube sizes, the resulting average velocity is in the neighborhood of 6 ft/sec. Applying the calculated tube area per stage to the number and size of tubes results in the length of tubes needed in the stage, and hence in the stage length for the longitudinal configuration of tubes. These calculations complete the condenser section design, as far as the preliminary design is concerned.

The dimensions of the brine channel are difficult to determine, the chief uncertainty being the degree to which vapor will escape from the flashing brine. The problem arises because the vapor formed due to expansion of the brine originates throughout the body of the flowing brine. In order for this vapor to escape from the free surface of the brine, it must either coalesce into large bubbles or be brought to the surface by turbulent motion of the flowing brine, the latter mode being the more important one at the temperatures in most of the stages. Several manufacturers of multiple-stage flash distillers have evolved systems of baffles downstream from the orifices to stir up the flashing brine and assist the vapor to escape. In spite of these provisions, it is observed that most flash stages operate with incomplete vapor discharge. The amount of incompleteness is measured by the nonequilibrium allowance, which is defined as the difference between the temperature of the flashing brine leaving the stage and the vapor saturation temperature corresponding to the stage pressure, plus the boiling point elevation. As pointed out by Porteous [14] and Best [15], the transfer of vapor from the water to the space above it is analogous to the transfer of heat, with temperature difference between the water and vapor being the driving force. Using this analogy, it is reasonable that there should be a terminal temperature difference between vapor and water, just as there is a terminal temperature difference for any heat exchanger. The existence of a terminal temperature difference means that the brine has not expanded all the way to the temperature corresponding to the vapor pressure. Complete expansion to that temperature would have produced the calculated amount of vapor, but since complete expansion is impossible, the actual vapor output of a stage will be slightly less than the equilibrium calculations would suggest.

Brine flow rates in existing plants vary from about 200,000 lb/hr to nearly 1,000,000 lb/hr per ft of channel width. Stage lengths in comparably sized plants have also varied between 5 and 15 ft, all of which indicates the wide divergence of design practices in different companies.

III. SUMMARY

Multiple-stage flash distillation for converting seawater to freshwater has been developed since the middle of the 1950s to become the type of plant most favored for large-scale production. In spite of many attempts to eliminate much of the heat-transfer tubing, these plants still largely depend upon the use of tubular condensers and heaters.

From thermodynamic reasoning it can be shown that for given conditions of temperatures and heat-transfer coefficients, there is a unique relationship between steam economy, brine exit temperature, area economy, and recovery ratio, and an indirect relationship between plant efficiency and the number of stages. The minimum number of stages required in an idealized flash distiller has been found to be only slightly greater than its steam economy value for low-efficient plants, but the number becomes increasingly greater than the steam economy value as the latter is increased. For instance, in ideal plants having steam economies of 11 or more pounds of distilled water per pound of steam, the minimum numbers of stages are generally about 50% more than the numerical values of steam economy.

Designers in the United States seem to favor the long-tube construction, in which a single set of condenser tubes extends through several stages. In contrast, designers in the United Kingdom seem to favor separate sets of tubes for each stage, arranged either horizontally or vertically and running in a direction perpendicular to the flow of the brine. The longitudinal arrangement requires much more ground area than does the transverse configuration, but seems to require less pumping power.

REFERENCES

1. N.A. El-Ramley, J.M. English, and J.W. McCutchan, The Multistage Flash Desalting Process – Its Commercial Applications, Rept. No. UCLA-Eng-7079, Aug., 1970, Univ. of California, Los Angeles, California, 125 pp.
2. M.S. Sachs, Desalting Plants Inventory, Rept. No. 1, Office of Saline Water, U.S. Dept. Interior, Jan. 1, 1968, 25 pp.
3. Universal Desalting Plant Design, 5 vol., Office of Saline Water, U.S. Dept. Interior, 1969.
4. D.F. Othmer, Evaporation for Desalination – Improved Multistage Flash (MSF) Processes, Desalination, 6 (No. 1), Apr., 1969, pp. 13-24.
5. C.N. Hodges et al., Solar Distillation Utilizing Multiple-effect Humidification, Res. Develop. Progr. Rept., No. 194, May, 1966, Office of Saline Water, U.S. Dept. Interior, 159 pp.
6. Burns and Roe, Inc., First Annual Rept., Saline Water Conversion Demonstration Plant, San Diego, California, Res. Develop. Progr. Rept., No. 102, Office of Saline Water, U.S. Dept. Interior, Feb., 1964, 85 pp.
7. Flash-Evaporator Plant in Malta, The Engineer (London), 224 (No. 5834), Nov. 17, 1967, pp. 643-646.
8. J.T. Chambers, Performance Characteristics of Sea Water Distillation Equipment, Inst. of Eng. Res. Rept. Ser. No. 75, Issue No. 19, Nov. 6, 1959, Univ. of California, Berkeley, California, 31 pp.
9. OSW Saline Water Conversion Report for 1967, Office of Saline Water, U.S. Dept. Interior, 1967.
10. L.H. MacLeod et al., The Optimum Temperature for the Operation of a Non-Scaling Multi-Stage Flash Evaporator Plant, Sea Water Conversion Laboratory Rept. No. 63-1, Jan., 1963, Univ. of California, Berkeley, California, 42 pp.
11. R.J. Easterly, Flash: An IBM-7000 Code for Computing Multi-Stage Flash Evaporator Plant Designs for the Desalination of Sea Water, ORNL TM-124, Oak Ridge Nat. Lab., Tennessee, Sept., 1965.
12. C.T. Mothershed, Orosef: A Fortran Code for Overall Design of the Multi-stage Flash Desalination Plant, ORNL TM-1627, Oct., 1966.

13. C. T. Mothershed, Orsef: A Fortran Code for the Calculation of Multi-stage Flash Evaporation Desalination Plant Designs, ORNL TM-1542, March, 1966.
14. A. Porteous, A Low Temperature Model for Equilibration Rates in Flashing Flow and its Design Implications, Desalination, 6 (No. 3), Aug., 1969, pp. 337-348.
15. T.R. Best, Design of a Flash Stage for a Multi-stage Flash Evaporator, Paper No. 71-WA/PID-2, Amer. Soc. Mech. Eng., Nov., 1971, 9 pp.

Chapter 8

VAPOR COMPRESSION DISTILLATION

I. USAGE OF THE PROCESS

The vapor compression distillation process is important because of its highly efficient usage of mechanical rather than heat energy. Plants employing this process are especially attractive for applications where steam is not readily available for other distillation processes. For example, there was extensive usage of vapor compression distillers for freshwater production during World War II on those naval vessels which were powered by diesel engines rather than by steam turbines. Similarly, the use of portable units driven by internal combustion engines greatly simplified freshwater production at advanced military bases and other remote water-short areas, for example, survey and construction sites of the petroleum industry.

Vapor compression plants are relatively new, having come into commercial usage as recently as the 1930s. The impetus of the wartime applications noted above led to its rapid development in moderately small sizes. The first large fixed installation took place in the 1950s, when the United States Air Force adopted units of this type for its air bases at Dhahran in Saudi Arabia, Port Etienne in Mauritania, and at Kindley Air Force Base in Bermuda. The third distillation pilot plant of the OSW series, designed to produce 1 million gallons per day, was a vapor compression plant located at Roswell, New Mexico. The entries in Table 8-1 detail some of the more significant fixed vapor

TABLE 8-1

Fixed Vapor Compression Installations

	No. of units	Capacity kgal per day		Year
		Total	Per unit	
Ascension Island (British)	1	30	30	1966
Ascension Island (British)	17	131	7.7	
Dhahran, Saudi Arabia	4	200	50	1955
Gibraltar	1	25	25	1957
Gibraltar	1			1960
Hill A. F. B., Utah	1	40	40	1964
Hofra, Libya	3	45	15	
Kindley Air Force Base, Bermuda	4	200	50	1954
Port Etienne, Span. Sahara	2	57.6	28.8	1954
Roswell, New Mexico	1	1000	1000	1963
Salina, Ecuador	2	100	50	1956
San Juan, Peru	2	57.6	28.8	1954
Tripoli, Libya	3	30	10	
Turks Islands, West Indies	4	30	7.5	
Villa Cisneros, Spanish Sahara	3	30	10	
Total	48	1975.2		

compression installations. No attempt has been made to inventory the very large number of portable and mobile units.

As indicated in Chapter 4, the vapor compression plant consists primarily of a condenser-evaporator and a vapor compressor. Vapor generated on the evaporating side of the heat-transfer surface is compressed and delivered to the condensing side of the same heat-transfer surface. The latent heats of evaporating and condensing fluids are very nearly equal, so that the energy required by the compressor is only that needed to offset the boiling point elevation on the evaporating side and to provide a small differential to ensure the desired flow of

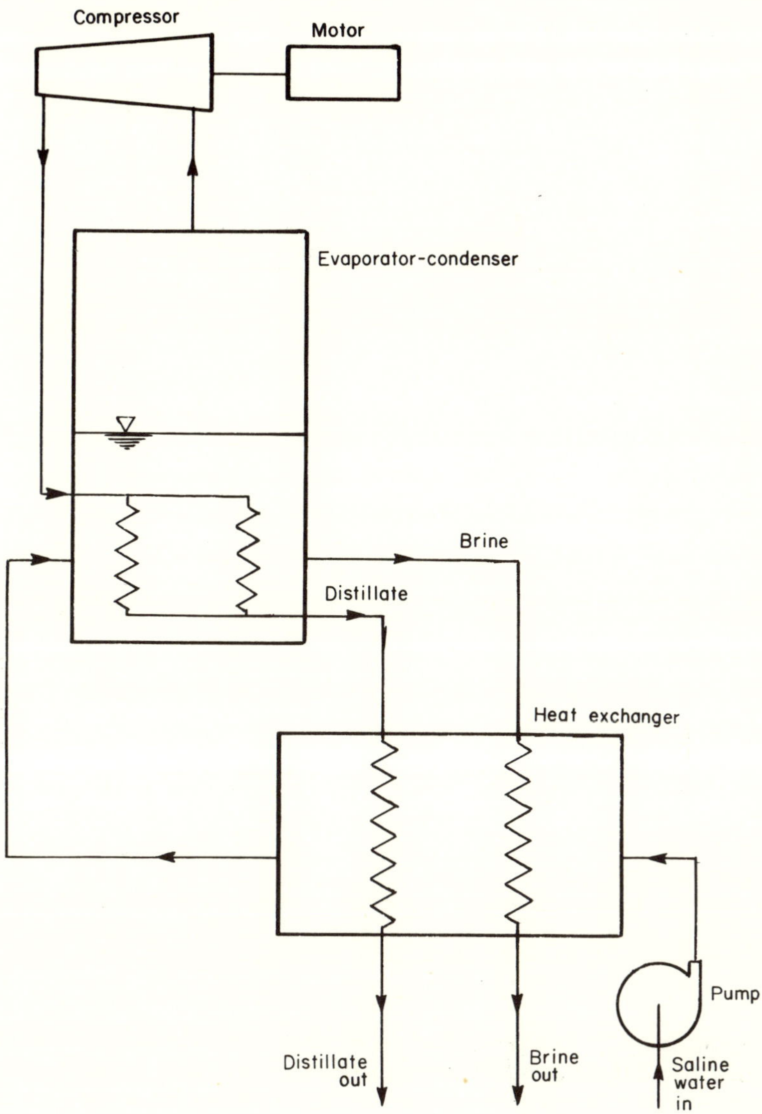

FIG. 8-1. Vapor compression distillation plant (schematic flow sheet).

heat. Since most vapor compression distillation plants to date have been designed to operate near the atmospheric boiling point, it is necessary to heat the incoming saline water to that boiling point before

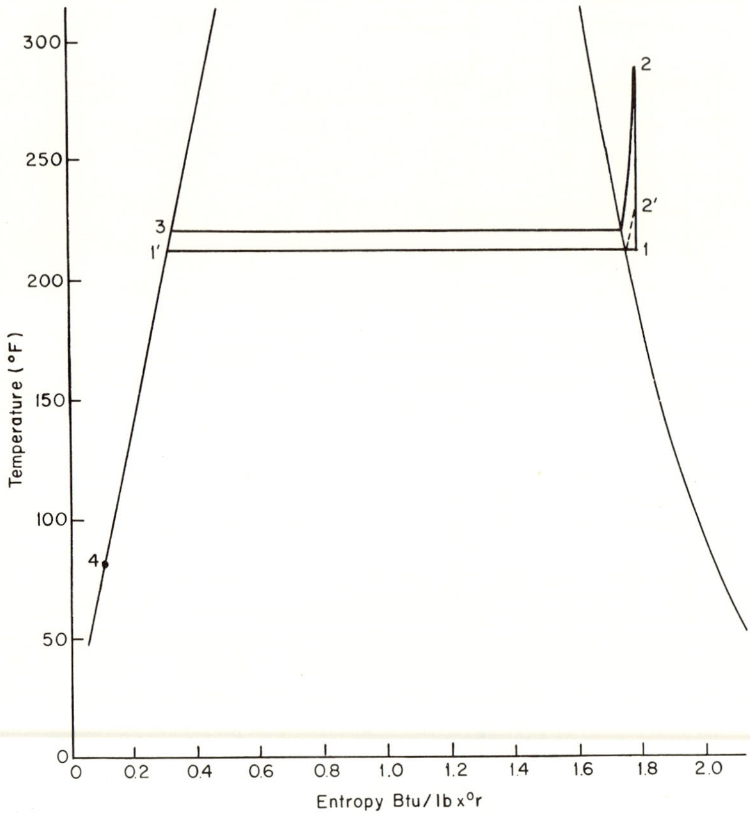

FIG. 8-2. Temperature-entropy chart of vapor compression distillation process.

evaporation. The heat required to accomplish this is obtained from cooling the outgoing brine and distillate, this heat transfer conventionally taking place in a liquid-to-liquid heat exchanger, as indicated in Fig. 8-1. The size and cost of these heat exchangers has become a deterrent to more extensive use of this type of plant. One scheme proposed to overcome this problem is the use of a multiple-stage flash distiller in place of the liquid-to-liquid heat exchanger, details of which appear in a later chapter.

II. THEORETICAL ANALYSIS

An approximation of the vapor compression distillation process is represented on the temperature-entropy diagram, shown in Fig. 8-2. Evaporation takes place along the line 1'-1, which is a line of constant temperature, and results in superheated vapor at statepoint 1. This vapor is compressed to statepoint 2 and is then passed to the heating jacket of the evaporator where it is condensed along the constant pressure line 2-3 to produce liquid at point 3. The liquid distillate then passes into the liquid-to-liquid heat exchanger where it is cooled to statepoint 4, and is finally discharged at this temperature. Meanwhile, the portion of the brine left after the vapor formation also passes into a liquid-to-liquid heat exchanger and is cooled from point 1' to point 4 before being discharged. The incoming saline water, entering the heat exchanger at a temperature less than that at 4, is heated to a temperature slightly less than at point 1'. The added heat required to change the temperature of the incoming water to the boiling point, 1', is derived from the condensation of a small portion of the vapor in the heating jacket of the evaporator. For the purpose of simple analysis, however, it will be assumed that the incoming saline water enters the evaporator at temperature 1'. The compression of the vapor is also assumed to be isentropic rather than isothermal, as in the ideal reversible case, because isothermal compression has not as yet been achieved in any practical device. It should be noted that compression from 1 to 2' is needed to overcome the boiling point elevation due to the salinity of the evaporating liquid, and that the further compression from 2' to 2 provides the temperature differential, t_3-t_1, for heat transfer in the evaporator-condenser.

A. Heat Balance Equations

The following equations can be written to express the interrelationships of energy.

Heat balance on the evaporator:

$$W(h_2-h_3) = W(h_1-h_1') \tag{1}$$

where W is the production rate and h is the enthalpy.

Heat balance on the liquid-to-liquid heat exchanger:

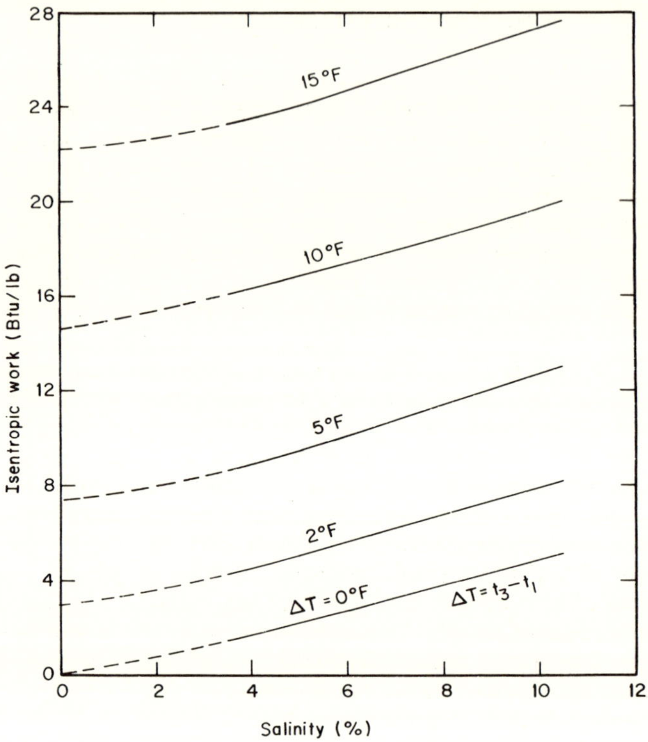

FIG. 8-3. Work of isentropic compression.

$$W_f c(t_1 - t_c) = Wc(t_3 - t_4) + W(W_f/W - 1)c(t_1 - t_4) \tag{2}$$

where W_f is the feed supply rate, t is the temperature, and c is the specific heat.

The overall energy balance is:

$$W(h_2 - h_1) = Wc(t_4 - t_c) + W(W_f/W - 1)c(t_4 - t_c) \tag{3}$$

Heat transfer in the evaporator:

$$W(h_2 - h_3) = UA\theta_m = UA(t_3 - t_1) \tag{4}$$

where U is the overall heat-transfer coefficient, A is the area, and θ_m is the mean temperature difference for heat transfer.

Heat transfer in the heat exchanger:

$$W_f c(t_1 - t_c) = U_1 A' \theta'_m \qquad (5)$$

If the specific heats are assumed to be equal, Eq. (2) can be solved for t_4, giving the result:

$$t_4 = t_c + \left(\frac{W}{W_f}\right)(t_3 - t_1) \qquad (2a)$$

Similarly, Eq. (3) can be manipulated to yield the equation:

$$t_4 = t_c + \frac{W}{W_f}\left(\frac{h_2 - h_1}{c}\right) \qquad (3a)$$

Comparing the last two equations gives the relationship:

$$c(t_3 - t_1) = h_2 - h_1 \qquad (6)$$

Equation (6) states that the energy input must equal the sensible heat involved in changing the temperature of water from that of evaporation to that of condensation.

B. Compressor Energy Requirements

The left-hand side of Eq. (3) represents the energy needed to compress the vapor. For given values of t_3 and t_1, the magnitude of (h_2-h_1) will vary with the concentration of the salts in the water being evaporated, increasing with the salinity. For instance, when salinity is doubled, as is customary in commercial plants, the increase amounts to approximately 1.4 Btu/lb. Values of (h_2-h_1) have been computed for several values of salinity and of temperature difference between evaporating and condensing vapor. These were plotted on Fig. 8-3, with values of the temperature difference for heat transfer, (t_3-t_1), as a parameter. An inspection of this graph will indicate the effect of salinity on the energy required for compression. For low values of (t_3-t_1), i.e., for high-efficiency plants, the added work of compression due to salinity represents a large fraction of the total energy required. While the figure applies to ideal reversible processes, the qualitative aspects of the curves indicate the relative changes in energy requirements of actual irreversible processes. For low values of (t_3-t_1) this represents a substantial fraction of the work of compression.

TABLE 8-2

Costs of Energy and Capital Amortization[a]

t_3	t_3-t_1	h_2	h_3	h_2-h_1	A	Energy cost	Capital cost	Sum of costs
217	5	1161.1	185.10	9.9	0.390	0.242	0.370	0.612
218	6	1162.5	186.11	11.3	0.326	0.276	0.309	0.585
219	7	1164.0	187.12	12.8	0.279	0.313	1.265	0.578
220	8	1165.4	188.13	14.2	0.244	0.347	0.232	0.579
221	9	1166.8	189.14	15.6	0.218	0.382	0.207	0.589
222	10	1168.3	190.15	17.05	0.196	0.417	0.186	0.603
223	11	1169.7	191.15	18.5	0.178	0.452	0.169	0.621
224	12	1172.6	193.13	21.4	0.151	0.523	0.143	0.666

[a] Figures basic to the above table are as follows: t_1 is 212°F; BPE is 1.6°F; h_1 is 1151.2 Btu/lb with entropy = 1.762 Btu/(lb × °R); U is 500 Btu/(hr × ft^2 × °F); η_{comp} is 70%. Capital cost is \$10. per ft^2, annual charges are 10%, and power cost is \$.007 per kWh.

C. Area Economy

Another factor having considerable effect on the cost of distilled water is the production rate per unit area, or area economy. The value representing this area economy may be obtained from Eq. (4) as follows:

$$\frac{W}{A} = \frac{U\theta_m}{h_2-h_3} = \frac{U(t_3-t_1)}{h_2-h_3} \qquad (4a)$$

For given values of U and (t_3-t_1), the area economy is dependent on the values of h_2, which in turn are determined by the salinity of the water being distilled. The higher the salinity of the water, the greater will be the value of h_2, and a consequent decrease in the value of area economy.

The deductions derived from Eq. (4a) refer to the surface area of the condenser-evaporator only, ignoring the area of the terminal heat exchanger. However, the cost of the terminal heat exchanger can be quite large, in spite of the fact it is called upon to transfer only about 1/10 as much heat as the evaporator. The relatively large cost is due to the small value of the heat-transfer coefficient characteristic of liquid-to-liquid heat exchangers, these coefficients lying between 1/4 and 1/8 of those found in the condenser-evaporators. Since the temperature difference for heat transfer in the terminal heat exchanger is nearly the same as that in the evaporator, it follows that for terminal heat exchangers with coefficients as low as 1/8 of those in the condenser-evaporator, the areas for heat transfer will be at least 8/10 of the latter, thereby nearly doubling the heat-transfer area cost.

Since it is observed that the cooling of the brine and condensate begins at 212°F, very nearly the same as the maximum temperature of the multiple-stage flash distillation process, the question has arisen as to whether the brine could be passed through a series of flash stages and there produce additional distillate, rather than being sent through the liquid-to-liquid heat exchanger, as described above. This leads to the concept of combination plants, which will be dealt with in a later chapter.

D. Series Operation to Offset the Effect of High Brine Salinity

An examination of Fig. 8-3 suggests that savings in compressor power might be achieved either by reducing the temperature difference across the heat exchanger or by reducing the brine salinity. A design

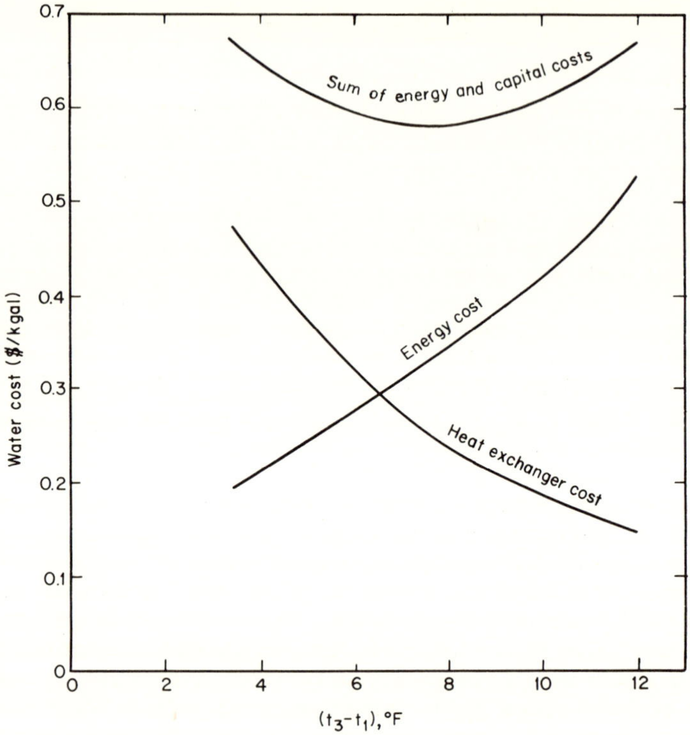

FIG. 8-4. Effect of heat-transfer differential temperature on cost of water. Conditions of plant: evaporation at 212°F, salinity of brine discharged is 7%, compressor efficiency of 70%, cost of evaporator \$10 per ft^2, cost of power \$.007 per kWh, and annual charges 10%.

involving a decreased heat-transfer temperature differential would require an additional heat-transfer surface area to offset that decrease, with the result that any costs accompanying that necessary area increase would negate the proposed compressor power saving. The other suggestion for power saving, namely, a reduction in brine salinity, would employ several vapor compression distillation plants, the brine circuits of which would be operated in series. In such an arrangement the seawater fed to the first unit would be only partially concentrated before it would flow to the next unit to be further concentrated, reaching its final maximum salinity only in the last unit. As a result, the power required in the units of lower salinities would contribute to a total power requirement less than the power that would be required to produce the same quantity of water in a single unit involving the maximum salinity.

Chambers and Larsen [1] ran tests to confirm the above theoretical conclusions, and showed that an appreciable saving was indeed possible through staging of the brine circuit. Their results indicated that two units, operating in series and doubling salinities, would show a power saving of about 10% as compared with a single unit producing the same amount of water and concentrating the brine to the same degree. This type of arrangement, operating in series to accomplish power saving by means of a reduction in brine salinities, would find its greatest justification in cases where large-scale water production would require more than a single unit to realize the necessary supply.

E. Relative Effect of Power and Area Economy on Cost of Distillation

It is readily recognized that compressor power requirements and area economy both have important effects on the cost of distilled water, but in opposite directions. For example, with an increase in the temperature difference (t_3-t_1) the compressor power must be increased for a given output, but the heat-transfer surface requirement may be reduced. The result of these opposite effects upon costs are variable, but will usually show a minimum total cost value for some particular temperature difference. To illustrate this point, a series of calculations for enthalpies and cost components were made for several values of t_3 and tabulated in Table 8-2, based on the assumed conditions listed in the table. These computed cost components for power and area are plotted on Fig. 8-4, with (t_3-t_1) as abscissal. Also plotted is the curve showing the sum of these components, which is seen to have a definite minimum value. While the choice of other conditions and values of parameters would lead to curves of a different set of values and to a different position of the minimum, Fig. 8-4 presents a correct qualitative picture of the relationships between the two cost components.

III. DEVELOPMENT TRENDS

As previously described, the two principal components of vapor compression plants are its vapor compressor and the condenser-evaporator. Attempts to design larger plants of this process type have necessitated the consideration of various means of increasing the capacity of the compressor and of improving the operation of the condenser-evaporator. Since these two objectives are somewhat independent of each other, the following paragraphs will attempt to consider the selection of the type of compressor that can be used, and the enhancement of heat-transfer surfaces to improve the operation of the condenser-evaporator.

A. Selection of the Type of Compressor

The type of compressor used in a vapor compression plant has been largely determined in the past by the method by which the plant attempted to meet the problem of scale control. The vapor compression process was invented and developed at a time when the standard method of controlling scale in other types of distillation plants was by the use of additives containing polyphosphates. Such a scheme could not be used for vapor compression plants because the maximum temperature at which polyphosphates are effective is 190°F, and the temperature of evaporation of the brine within these early vapor compression plants was 212°F or more. The reason that these plants operated at this high temperature was due to the decision to use atmospheric pressure in the evaporator rather than a lesser pressure, in order to avoid the expense and complication of vacuum equipment.

Because the early plants were operated without pretreatment of the feedwater for alkaline scale control, it was inevitable that scale formation on the heat-transfer surfaces would present serious problems. The thickness of scale increased as operation of the plant continued, imposing additional and increasing resistance to heat flow, which in due time tended to reduce the plant capacity for producing distilled water. To offset this and to maintain the capacity nearly constant, direct displacement compressors were installed. These were generally of the gear or Roots type since they would permit the increase of discharge pressure without appreciable change in the volume of vapor being compressed. In the operation of these plants, the power input would gradually be increased as the scale built up, until some predetermined maximum permissible amount of power input was reached, at which time the plant would shut down to allow the scale to be removed by the circulation of an acid rinse over the heat-transfer surface. Operation would then be resumed and the same sequence of events be repeated.

The use of direct displacement compressors, as described above, placed a capacity limitation on vapor compression plants, since these compressors were limited in their volumetric displacement to a size which corresponds to about 50,000 gpd of water production. Other types of compressors, such as centrifugal and axial flow machines could be constructed in much larger sizes, but they have characteristics such that increases in pressure are accompanied by appreciable decreases in volumetric capacity. Because such a decrease would be unacceptable, considerable investigative effort has been directed to the solving of this problem.

The most successful solution to the above problem is the application of acid pretreatment to the feedwater to eliminate alkaline scale formation. With this danger minimized, both axial flow and centrifugal compressors can become practical, thereby permitting vapor compression plants to be designed for much larger capacities. For example, the OSW at Roswell, New Mexico [2], utilizing an axial flow compressor, was successfully designed to furnish distilled water at the rate of one million gallons a day. Because of the pretreatment of feedwater, the plant encountered no difficulties with alkaline scale, although did experience problems with calcium sulfate scale because of the exceptionally large percentage of that salt in the ground water which served as feed to the plant.

A variation of the usual acid treatment for scale control has been introduced by Williams [3] in which a "slug" of acid is fed to the feedwater stream at stated intervals. A plunger pump, mounted on the side of the feedline near the heat exchanger is filled with a concentrated acid solution, and is manually pumped to force the contents of the plunger cylinder into the feedstream at intervals of 6 to 8 hr. This method has found application in United States naval installations, where mechanics are in attendance at all times, and has been found to effectively dissolve any scale which might have been deposited during the previous time interval.

B. Heat-Transfer Enhancement

Numerous attempts have been made to reduce the resistance to heat flow through the solid wall separating the distillate from the evaporating brine. The most fruitful of these achievements has been to improve forced circulation within the tubes and to develop various techniques for limiting the amount of saline water covering the evaporating surface to a thickness of a very thin film.

Forced and positive circulation of water over a metallic heat-transfer surface has been productive in improving heat transfer in many evaporators. The OSW vapor compression plant at Roswell, New Mexico, used forced circulation in vertical tubes to increase the heat-transfer coefficient of the evaporating water. The design of this plant was based on the thesis advanced by Dodge and Eshaya [4], that substantial improvement in heat-transfer rates on the evaporating side

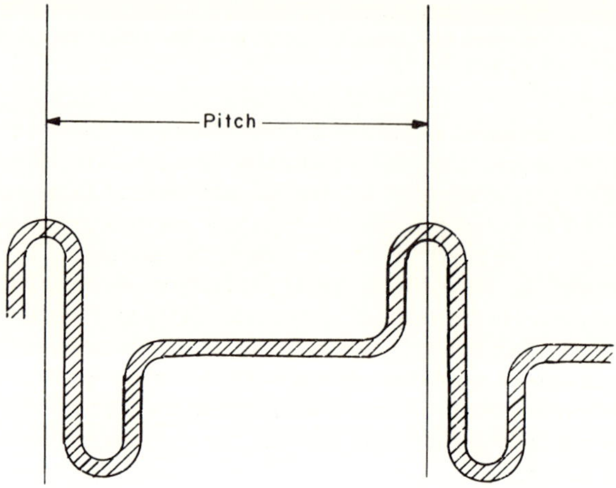

FIG. 8-5. Enhanced heat-transfer surface. [After Thomas & Alexander [7].] Note that pitch in 2 in. diameter experimental tubes was 1/8 in. using 0.004 in. thick stainless steel.

of the heat-transfer surface was an essential requirement to reduce the cost of distilled water. Their reasoning was based on the assumption that the heat-transfer coefficient was subject to the Dittus-Boelter [5] equation:

$$hD/k = 0.023$$

where h is the heat-transfer coefficient, D is the tube diameter, k is the thermal conductivity, Re is the Reynold's number, and Pr is the Prandtl's number. From this equation the heat-transfer coefficient is seen to be proportional to the 0.8 power of the velocity. Thus, they reasoned that increasing the velocity through the tubes would increase the heat-transfer coefficient almost proportionally. However, the increase in velocity could be accomplished only at the expense of increased energy consumed the pipe friction, so that the overall economy was not as great as anticipated.

Another important development in heat-transfer enhancement was the use of the fluted vertical tube. Introduced in the desalting industry by Lustenader et al. [6], it has been investigated and improved, so that its performance can be rather accurately predicted. Physically, the condensation of water on a fluted surface results in the drainage of condensate into and along the concave portion of the flutes, thereby keeping the convex portions dry, so that there is direct contact between the cold dry surface and the vapor to be condensed. The condensate film which forms on a smooth surface and impedes heat flow between a cold surface and vapor is thus limited to the concave portions of the grooves. Grooves on the evaporating side of the heat-transfer surface can also show improved heat-transfer coefficients. The mechanism by which the improvement in heat transfer is accomplished in this case is not well understood, but the improvement is such that double-fluted tubes are distinctly advantageous so far as heat transfer is concerned. The arrangement of flow in such systems has been to use the outside of the tube for condensation and the inside for evaporation. With downward flow of the brine, the liquid is either sprayed against the upper portion of the tube interior, or is admitted by flowing over a weir formed by the extended periphery of the upper end of the tube.

The shape of flutes has been the subject of investigation by Thomas and Alexander [7]. They reasoned that simple double flutes, produced by deforming a thin-walled tube, would have the disadvantage that the thin portion of the water film on one side of the surface would be opposite the thick portion of the water film on the opposite side. By a special shaping of the fluted surface, as in Fig. 8-5, they were able to place the thin-film portion of the evaporating surface opposite the thin-film portion of the condensing surface. They experimented with tubes of many sizes and shapes. Their published conclusions of these tests with 2 in. tubes indicated that overall heat-transfer coefficients for temperature differences of 5°F were about 1650 Btu/(hr $\times$ ft^2 $\times$ °F) for simple corrugations, 3300 for the modified form of Fig. 8-5, and 550 for smooth tubes.

Another means of producing thin films is the use of wiper blades held against the surfaces moving relative to the blades. Lustenader et al. [6] introduced this concept also, using a fixed vertical tube with wiper blades rotating inside the tube. Water admitted ahead of the wiper was spread smoothly across the surface. While this scheme somewhat improved heat transfer, the mechanical complication prevented its practical application. More recently Tleimat [8] has reported the use of stationary radial wipers held against the flat surfaces of rotating disks. The disks were arranged in pairs and sealed over narrow rings at the periphery to form a flat cylinder. Water, evaporated from the wiped films on the outer ends of the cylinder, was compressed and condensed on the inside of the cylinder ends. Centrifugal forces due to rotation kept the internal condensate film thin by throwing the water outward. The pool of condensate, formed at the periphery, was picked up by the stationary scoops and drained through the hollow shaft. Overall heat-transfer coefficients amounted to approximately 2800 Btu/(hr $\times$ ft^2 $\times$ °F) at low rotor speeds and a temperature difference of 5°F. By increasing the speed by a factor of three, the reported results showed overall transfer factors of nearly 7000 Btu/(hr $\times$ ft^2 $\times$ °F) with a temperature difference of only 2°F.

Another means of increasing heat-transfer coefficients through the production of thin films of liquid on heat-transfer surfaces was that of using centrifugal force to cause the water to flow radially outward on the surface of the dished circular plates. Introduced for vapor compression by Hickman [9], this scheme involved a series of these dished circular plates mounted on a vertical shaft and arranged so that the saline water feed would be sprayed onto the concave surface of each plate, and the compressed vapor would be condensed on the opposite or convex surface. This was accomplished by mounting the plates in pairs with concave surfaces facing each other and sealing the edges of each pair around the periphery. Holes in the centers of each plate, except the lower plate of the bottom pair, provided a passage for the vapor to rise along the shaft to the compressor inlet at the top of the several pairs of plates. Adjacent pairs were sealed together around the periphery of the center hole to provide the vapor channel. Saline water feed, sprayed on the inner surface of each pair, was partially evaporated, and the residue of saline concentrate was collected by stationary scoops and conveyed to the outside of the enclosure. The entire rotating assembly was housed in a vertical cylinder, thus confining the compressed vapor leaving the compressor, and permitting it to come into contact with the convex surfaces of the rotating plates, where it condensed. The vapor condensing on these plates as water was thrown out by centrifugal force, collected in gutters attached to the wall of the stationary cylinder, and conveyed to the outside of the enclosure. The most developed form of this equipment was of a size appropriate for use by a single family and as a marketable item has not as yet proved successful.

REFERENCES

1. J. T. Chambers and P. S. Larsen, Series Staging of Vapor Compression Distillation, Series 75, Issue No. 20, Sea Water Conversion Lab., University of California, Berkeley, California.
2. P. L. Geiringer et al., First Annual Report: Brackish Water Conversion Demonstration Plant No. 4, Roswell, New Mexico, Res. Develop. Prog. Rept., No. 169, Office of Saline Water, U. S. Dept. Interior, 1965, 87 pp.
3. J. S. Williams, Saline water conversion for Advanced Naval Bases, National Academy of Sciences—National Research Council Publication No. 568, Sec. 6, 1958.
4. B. F. Dodge and A. M. Eshaya, Economic Evaluation Study of Distillation of Saline Water by Means of Forced-Circulation Vapor-Compression Distillation Equipment, Res. Develop. Prog. Rept. No. 21, Office of Saline Water, U. S. Dept. Interior, 1958, 122 pp.

5. F. W. Dittus, and L. M. K. Boelter, University of California Publ. Engineering, 2, 1930, p. 443.
6. E. L. Lustenader, et. al., The Use of Thin Films for Increasing Evaporation and Condensation Rates in Process Equipment, J. Heat Transfer, 81, 1959, pp. 297-307.
7. D. G. Thomas and L. G. Alexander, Imposed High-Performance fluted Tube for Thin-Film Evaporation and Condensation, Desalination, 8 (No. 1), July, 1970, pp. 13-20.
8. B. W. Tleimat, Condensation and Evaporation from a Rotating Flat-disk Wiped Film Evaporator, Sea Water Conversion Lab., Rep. No. 69-3, Univ. of California, Berkeley, California.
9. K. C. D. Hickman, et al., Centrifugal Phase-barrier Recompression Distillation, Advan. Chem. Ser. No. 27, pp. 128 and 146, 1960.

Chapter 9

COMBINATION DISTILLATION PLANTS

Each of the three types of distillation plants described in the previous chapters has certain advantages as compared with the other two. Consequently, combinations of the three have been proposed in an effort to gain the combined advantages of two or more types. The multiple-effect plant has the advantage of producing vapor from brines of varying degrees of salinity in which the concentration in the first, or hottest, effect is only slightly greater than the feed stream. In this respect, multiple-effect distillation is superior to the other two, since multiple-stage flash plants produce vapor from brine which is recirculated, and therefore is initially much more concentrated than the feed stream. Similarly, vapor compression distillation produces vapor from a brine of maximum concentration. Since the boiling point elevation increases with salinity, the average boiling point elevation during vapor formation is less for the multiple-effect plant than for the other two and the loss due to such elevation is correspondingly less.

The multiple-stage flash distillation plant has the advantage that the incoming mixture of saline feed and recirculated brine is heated without evaporation and flashing evaporation occurs without heat transfer, thus reducing the tendency for scale to form on the heat-transfer surfaces. In the other two types of distillation the vapor is formed during the transfer of heat, so that the saline water is concentrated locally at points where vapor is produced. This concentration leads to increased tendencies for scale to form.

The vapor compression plant has the advantage of recirculation of heat required for vapor formation, so that the amount of heat which must be dissipated from the plant is much less than the latent heat of the vapor formed. In contrast, both multiple-effect and multiple-stage flash plants require the dissipation of heat equivalent to that added in the first effect and brine heater, respectively. This quantity is much more than the heat equivalent of the compressor energy requirement of the vapor compression plant.

The advantages of the three types of plants have been utilized in several different combinations, the most obvious being:

Combination of multiple effect with multiple-stage flash
Combination of multiple effect with vapor compression
Combination of multiple-stage flash with vapor compression
Combination of multiple effect with both multiple-stage flash and vapor compression

Plants embodying these combinations have been either constructed for production and test purposes, or have been the subjects of extensive studies and design efforts.

I. COMBINATION 1: MULTIPLE EFFECT WITH MULTIPLE-STAGE FLASH

This combination is exemplified by the Clair Engle Distillation plant in San Diego, California [1], a demonstration plant of the OSW in which its three effects each contain a number of flash stages. The diagram of Fig. 9-1 shows the arrangement of effects and stages. It will be observed that the first effect is supplied with heat from an external source, but that the second and third effects receive their heat energy from the next higher effect, the amount of such heat being the heat rejected from the preceding effect. This particular plant operates with a wide range of temperature and hence shows a high value for its GOR.

Another possible manner of combining the multiple-effect and multiple-stage flash principles is to arrange it so that vapor is produced both by flashing and by heat transfer in each effect. An example of this version is the invention of Bromley [2], referred to by him as the MEF, or the multiple-effect flash plant. In this scheme the several effects are arranged in a vertical tower with the high-temperature ef-

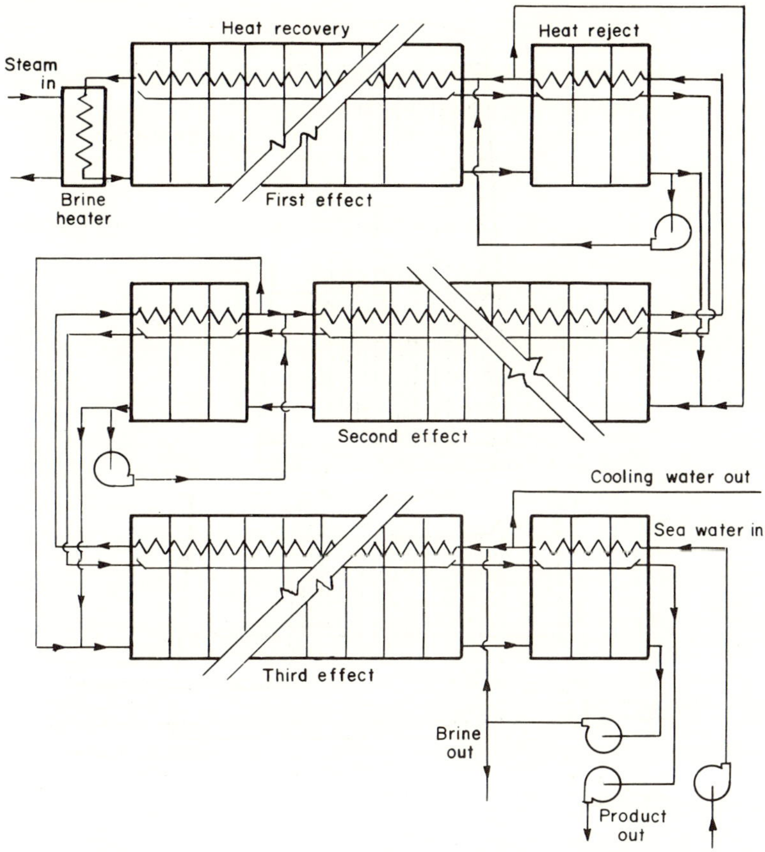

FIG. 9-1. Combination multiple-effect and multiple-stage flash distillation plant (schematic only). The Clair Engle Plant at San Diego uses this general type of flow sheet. Note that each effect is a complete multi-stage flash distiller. Pretreatment and vacuum systems are not shown.

fect at the top, as shown in Fig. 9-2. In this figure it is seen that the deaerated feed is introduced at the bottom of the tower and rises through condenser tubes to the top. During passage through these tubes the feed is heated by condensation of part of the steam produced in the several effects. The heated feed enters the evaporating tubes of the first effect through orifices, where a small amount of vapor is produced by flashing, but it is also heated by external steam condensing on the outside walls of its evaporating tubes. This heating causes

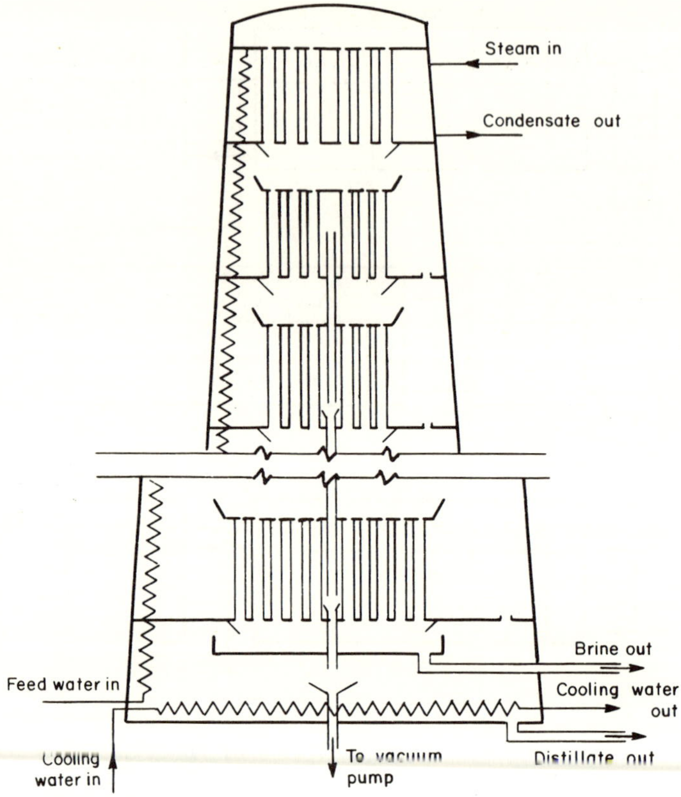

FIG. 9-2. Multiple-effect flash evaporator after Bromley [2]. (Schematic only.) Note the restrictions at the entrance of each tube, which cause flashing formation of vapor. Condensed distillate cascades from stage to stage through control orifices. Pumps, vacuum, and pretreatment systems are not shown.

the formation of additional vapor in this effect, separating itself from the liquid which enters the evaporating tubes of the second effect and flashes as it passes through orifices at the entrance to the tubes. Vapor formed by this flashing cools the liquid in the tubes, which in turn condenses the vapor from the preceding effect on the outside walls of the evaporating tubes of the second effect. The distillate thus formed in each effect is discharged through orifices to the next effect where it forms vapor by flashing and augments the heating in that effect.

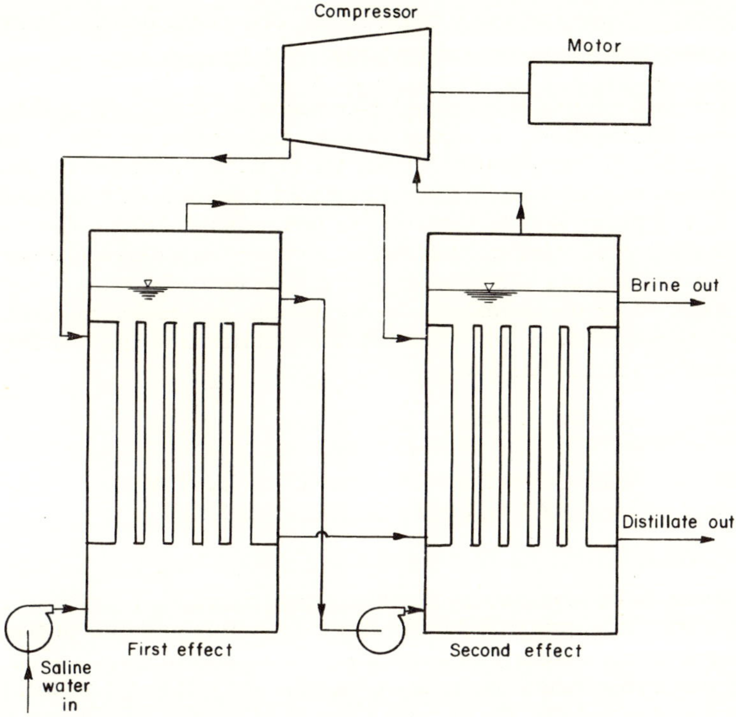

FIG. 9-3. Combination vapor compression and multiple-effect distillation plant after Roswell [3]. (Schematic only.) Note that forced circulation of water is upward through the evaporator tubes. Feed pretreatment system is not shown.

In the above plant it will be noted that the arrangement of condenser tubes and of distillate collection is quite similar to that of the multiple-stage flash plant. In the MEF plant the brine heater produces vapor, rather than being used only for increasing the temperature of the feed-water, as in the multiple-stage flash plant. The brine, cascading through the tubes of the several effects, performs similarly to the brine in a multiple-effect plant. The plant shown operates as a single pass unit, without recirculation of the brine as in a conventional multiple-stage flash plant. While there are as yet no prototype plants of the MEF type, there are indications that plants of this type may be built.

II. COMBINATION 2: MULTIPLE EFFECT WITH VAPOR COMPRESSION

The combination of multiple-effect distillation with vapor compression distillation is exemplified by the 1 mgd plant at Roxwell, New

Mexico [3]. Constructed by the OSW as one of its demonstration plants, this plant distills saline water of about 20,000 ppm initial salinity. The plant arrangement is shown in Fig. 9-3, and is seen to consist of two effects with a vapor compressor taking vapor from the low-temperature effect, compressing it, and delivering the compressed vapor to the high-temperature effect for heating that effect. This combination has the advantage that only about one-half of the vapor produced in the plant passes through the compressor. Since the compressor is a very expensive component in a vapor compression plant, this saving in compressor size represents a substantial savings in capital investment. Another advantage is the saving in cooling needed, since the vapor compression principle leads to minimizing the amount of heat which must be dissipated.

III. COMBINATION 3: MULTIPLE-STAGE FLASH WITH VAPOR COMPRESSION

This combination has been frequently proposed in order to obviate the need for the liquid-to-liquid heat exchangers implied in the ordinary vapor compression plant layouts. This combination seems particularly desirable since the range of temperature through which the outgoing brine and distillate must be cooled in a vapor compression plant is approximately the same as that required for the production of vapor in a multiple-stage flash plant. It seems logical that this cooling range should be used so as to produce additional distillate rather than being used merely to heat the incoming liquid feed.

An example of the use of a multiple-stage flash distiller for the preheating of water in a vapor compression distillation plant is that proposed by Tleimat [4], in which the MSF unit is employed to heat seawater from 70 to 220°F. Further heating of the feed to 228.5°F is accomplished by the condensation of steam from an external source in a feed heater. The hot feed then passes through five evaporators in series, each evaporator having a vapor compressor to remove vapor from the brine compartment and deliver it to the condensing side of the heat-transfer surface. Figure 9-4 illustrates the principle involved and is simplified from the Tleimat scheme by showing a single evaporator and compressor rather than five. The calculations presented in reference [4] indicated that a plant of this type producing 26.65 mgd would cost about $0.52 per daily gallon of capacity, and yield water costing 25¢/kgal. Approximately 5/6 of the water would be produced in the vapor compression section.

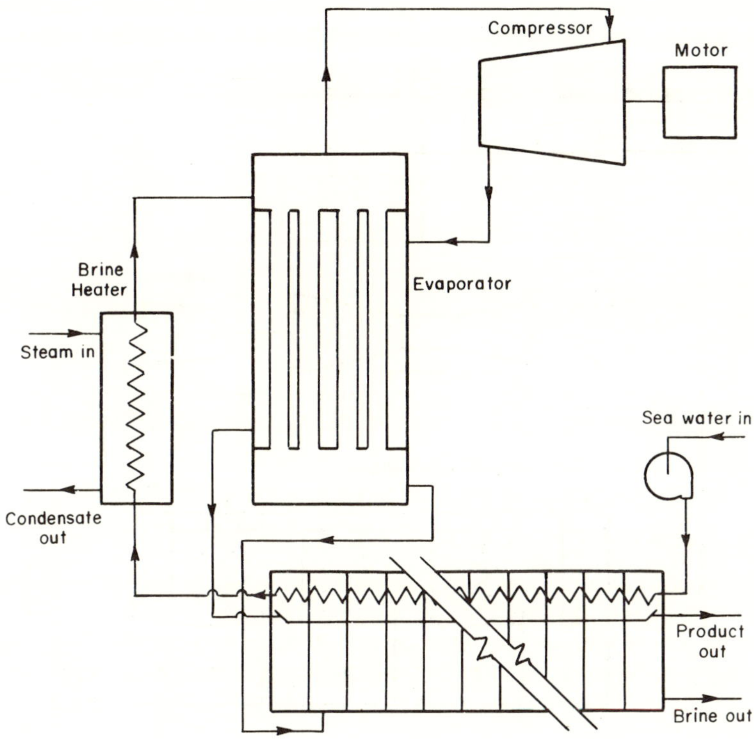

FIG. 9-4. Combination vapor compression and multiple-stage flash distillation plant adapted from Tleimat [4]. (Schematic only.) Note that feed pretreatment and vacuum systems are not shown.

IV. COMBINATION 4: MULTIPLE EFFECT WITH BOTH MULTIPLE-STAGE FLASH AND VAPOR COMPRESSION

This combination has been the subject of a number of studies, one example being the 8 mgd plant proposed by the Struthers Co. [5]. In this report a design is presented in which seawater is first heated in a 24-stage flash distiller and then passed to a four-effect distiller. Vapor from the fourth effect is compressed and delivered to the first effect as heating steam. The exhaust gases from the gas turbine used to drive the vapor compressor are used to produce steam in a boiler, this steam being delivered to a backpressure turbine for generating the electrical power required in the plant. The steam exhausted from the turbine is used to supplement the heat supply to the first effect,

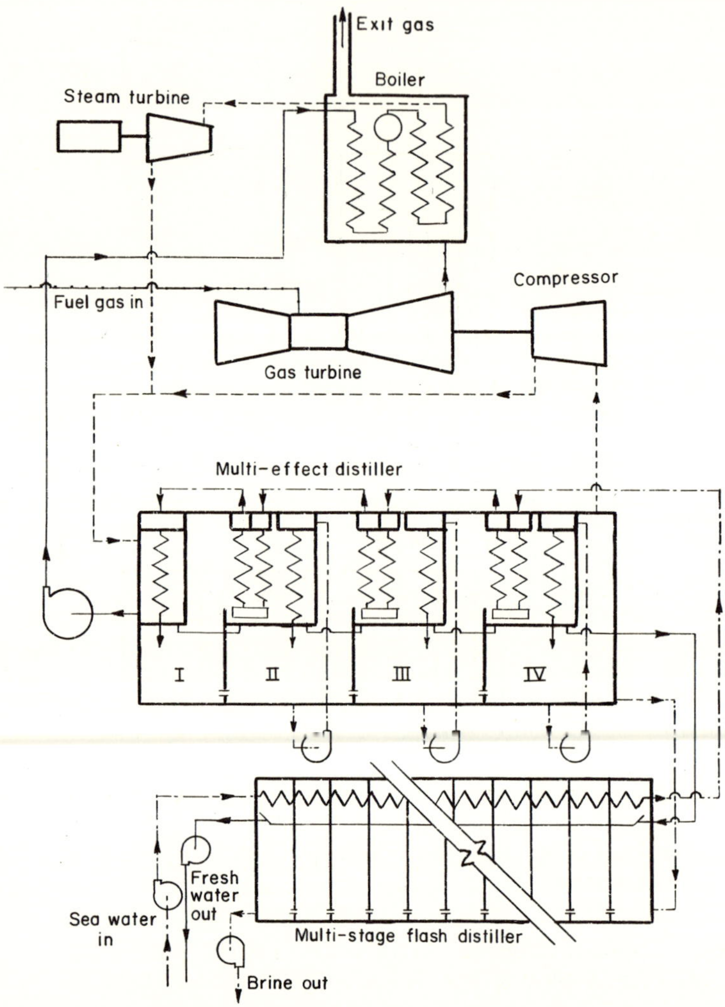

FIG. 9-5. Combination multiple-effect, multiple-stage flash and vapor compression distillation plant after Struthers [5]. (Schematic only.) Note that the proposed 8-mgd plant is driven by gas turbine and that waste heat from this turbine drives turbogenerator and furnishes part of the steam input to the multi-effect distiller. ——, Distillate; —·—, seawater and brine; ---, vapor; —·—·—, gas.

contributing a little less than 10% of the heat. The hot brine and distillate leaving the fourth effect are cooled by evaporation in the MSF plant where the feedwater is being heated. Figure 9-5 shows a simplified diagram of the plant, and illustrates the relationship of the three component plants. The design presented in reference [5] indicates an estimated PR of over 19 lb of distillate per 1000 Btu, with the MSF distiller producing slightly more than 10% of the product. The low production contribution of the MSF distiller is due to the fact that there is no recirculation in the MSF plant and the total feed to the plant is only slightly more than twice the production rate.

The above descriptions pertain to only a few of the possible configurations of combined plants, and are offered as suggestive of the range of such combinations. It seems clear that the use of simple plants of any of the separate types is less economical than some combination of the different types. It is therefore to be expected that the future will see more combination plants than single type plants, thus changing the pattern followed to date in which the single types have dominated.

REFERENCES

1. OSW Saline Water Conversion Report for 1967, Office of Saline Water, U.S. Dept. Interior, 1967.
2. L.A. Bromley and S.M. Read, Multiple-effect flash (MEF) evaporator, Desalination, 7 (No. 3), Apr. 1970, pp. 343-391.
3. American Hydrotherm Corp., First annual report, Brackish Water Conversion Demonstration Plant No. 4, Roswell, New Mexico, Office of Saline Water, US Dept. of the Interior, Res. Develop. Progr. Rep. No. 169.
4. B.W. Tleimat, Novel Approach to Distillation by Vapor Compression Distillation, Preprint Paper 69-WA/PID-1, A.S.M.E., Nov. 1969, 12 pp.
5. Struthers Energy Systems, Inc., Design and Economic Study of a Gas Turbine Powered Vapor Compression Plant for Evaporation of Sea Water, Res. Develop. Progr. Rept. No. 377, Office of Saline Water, U.S. Dept. Interior.

Chapter 10

DISTILLATION USING NONFUEL SOURCES OF ENERGY

The cost of energy has been shown to be a major component of the cost of distillation when fuel is its source. The question therefore arises as to whether such costs could be lowered by using energy from nonfuel sources, such as solar energy, waste heat, and geothermal energy. The devices appropriate for these sources of energy are considered in this separate chapter since their characteristics are quite different from those using conventional fuel. Solar energy is a low-intensity source, is most readily utilized at low temperatures, but can be concentrated to give high temperatures at some additional cost. Waste heat is available at various temperatures and, at high temperatures, would serve the conventional distillers of the previous chapters. At low temperatures it can be used in solar-type units. Geothermal heat is available at a limited number of locations in either hot-water or vapor forms, depending on very local conditions. These liquids and vapors from deep in the earth's crust contain dissolved materials quite different from those found in saline ground waters or in seawater, and therefore require special treatment.

I. SOLAR ENERGY FOR DISTILLATION

A. The Nature of Solar Energy

It is generally agreed that the sun's energy originates from nuclear fusion reactions occurring deep inside of a huge mass of gas, and that the quantity of radioactive gases in the sun is sufficient to support the fusion reaction for a billion years or more. This energy is dissipated very much as if it were black-body radiation from the outside of a gaseous envelope. The intensity of this radiant energy in the vicinity of the earth is essentially constant with time, but at the earth's surface its measured intensity varies with latitude, season, and atmospheric clarity. For any given location this radiation intensity reaches a maximum in summer and decreases to a minimum in the winter. Furthermore, its diurnal variation is an important characteristic, shrinking to zero after sunset, remaining zero until sunrise, and then increasing to a maximum about midday.

1. Intensity of Solar Energy

The maximum intensity of solar radiation is about 400 Btu/ft^2/hr on a surface perpendicular to the sun's rays at midday. Devices for collecting solar energy must be capable of absorbing this maximum intensity, but must also be designed to utilize as effectively as possible the lesser solar radiation intensities at other hours and at other angles. Since solar stills are built with horizontal surfaces which are stationary and do not follow the sun, the maximum energy which could be absorbed is that which impinges on the stationary horizontal surface. The daily magnitude of this quantity is about 2500 Btu/ft^2 during the sunny portion of a clear day in summer at latitude 30°. In winter this diurnal quantity shrinks to 1200 Btu/ft^2 on clear days, and to much less on cloudy days.

Pyrheliometric data have been assembled for clear days for many areas of the earth, and the average values of these data for the entire year vary from 1500 to over 2000 Btu/ft^2/day in arid areas. While these values may seem low when compared with heat-transfer rates in the evaporators previously considered, it must be remembered that any solar radiation of intensities comparable to those requirements would destroy terrestrial life as it presently exists.

2. Effects of Weather on Solar Radiation

For any particular location it is highly desirable to have pyrheliometric data in order to evaluate the effects of weather on solar radiation intensity. Lacking such data, it is possible to make first approximations based on published data of general weather conditions. The maximum energy available to any solar device will depend obviously upon the weather characteristics of a "clear day." Figure 10-1, plotted from data by Houghton [1], gives the diurnal incident radiation on clear days for several latitudes. These data were obtained by heat balance methods and may be regarded as conservative. Since Fig. 10-1 gives the total radiation for the day, it will be necessary in the case of most calculations to distribute this total radiation among the several hours of sunshine. This can be done most accurately by the use of pyrheliometric data for a specific geographical location, and from these data it will be observed that the instantaneous solar intensity on a horizontal surface varies between sunrise and sunset almost sinusoidally. When pyrheliometric data are not available, various schemes for estimating hourly radiation intensity have been proposed. The charts formulated by Whillier [2] illustrate one such scheme.

The wide variation of solar radiation intensity, both diurnally and annually, requires that any device for using solar energy to distill water must be capable of operating at widely varying temperatures. Because of this it is obvious that the variable intensity of solar energy produces a very different set of problems from that of fuel heat. The solar distillation process fluctuates with the solar energy intensity, its production varying from zero for most of the night to a maximum in the early afternoon of a sunny day. Also the daily production rate varies by a factor of at least 4 from its minimum in winter to its maximum in summer. Any operation using this source of energy presents a definite contrast to the constant production rates required in plants using fossil fuels.

B. Solar Stills

1. Basin-Type Solar Stills

This type is the only still which to date has been successfully applied to water distillation. It is the application of the natural hydrologic cycle to a relatively small apparatus. As noted in a

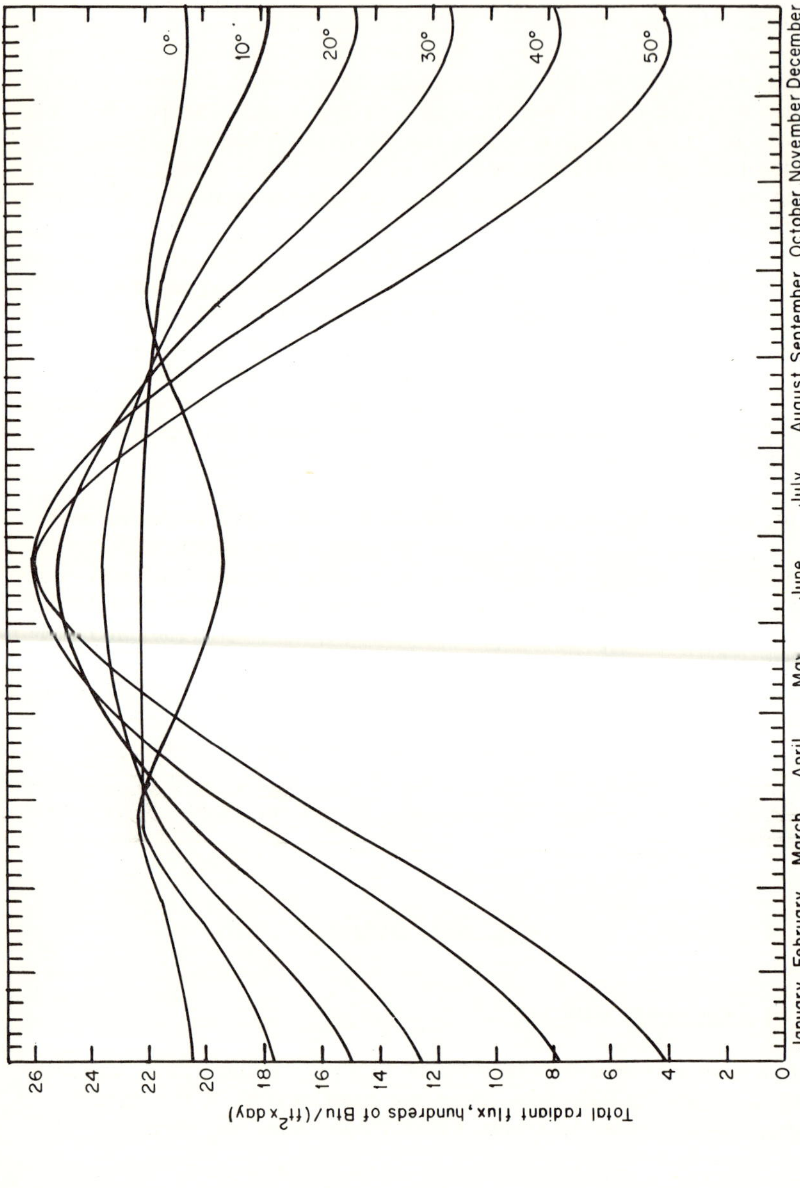

FIG. 10-1. Total diurnal radiant flux on a horizontal plane at the surface of the earth. (Numbers on curves are north latitudes in degrees.) Plotted from data of Houghton [1].

previous chapter, the essential features of the natural hydrological cycle are the production of water vapor above the ocean water's surface, the transportation of this vapor to cool regions by wind, and its subsequent condensation. The basin-type solar still reproduces this sequence of events.

A shallow basin of water is enclosed by a canopy which is transparent to solar radiation. The canopy is generally supported so that it slopes toward the sides of the basin and terminates at the condensate collection troughs. An approximate cross section of such a solar still is shown in Fig. 10-2. The basin is blackened on the surface to ensure maximum absorption of solar radiation for effective heating of the water. The transparent and vapor-tight enclosure serves to permit radiation to enter and to prevent escape of the humid air, to act as a radiation shield in reducing energy loss due to long-wave radiation from the water surface, and to provide a relatively cool surface on which part of the humidity can condense. Glass or certain sheet plastic materials are used for this transparent canopy. While these substances are not completely transparent to solar radiation, they absorb and reflect only a very small fraction of it and hence are heated much less than the blackened water basin. It is because of the transparent characteristic of the glass and plastic materials that the inner surface of the transparent enclosure is sufficiently cool to serve as the condenser. In this manner all three features of the natural hydrological process are duplicated inside of the basin-type still.

Existing solar stills differ from each other in materials used and in geometry, but all include the elements shown in Fig. 10-2, namely, the water basin, the sloping transparent cover, and the distillate collection troughs. The water basin must be watertight and is usually lined with an impervious opaque material. The transparent cover must be sealed around the edges to prevent vapor loss and must be sloped at an angle sufficiently great to ensure that condensate formed on it will flow by gravity to the condensate troughs, rather than to drip back into the basin. The condensate troughs must be so arranged that they collect all of the condensate dripping from the lower edges of the canopy and convey this condensate to the outside of the enclosure. All of the large solar stills are built on the ground, and the basin liner laid directly on the ground or on some insulating material to reduce heat losses. Materials used for the structural elements have been wood, concrete, metals, and plastics.

The solar-still installation must include provisions for supplying saline water, discharging brine, and storing the distillate produced.

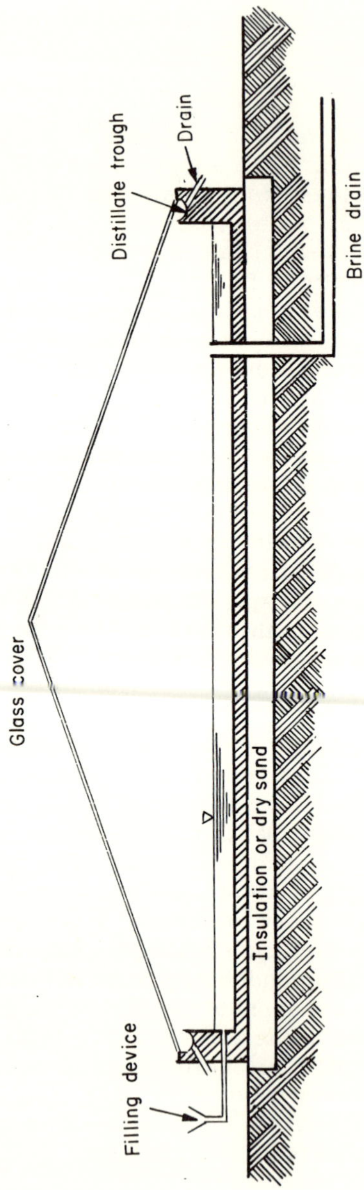

FIG. 10-2. Schematic cross section of basin-type still.

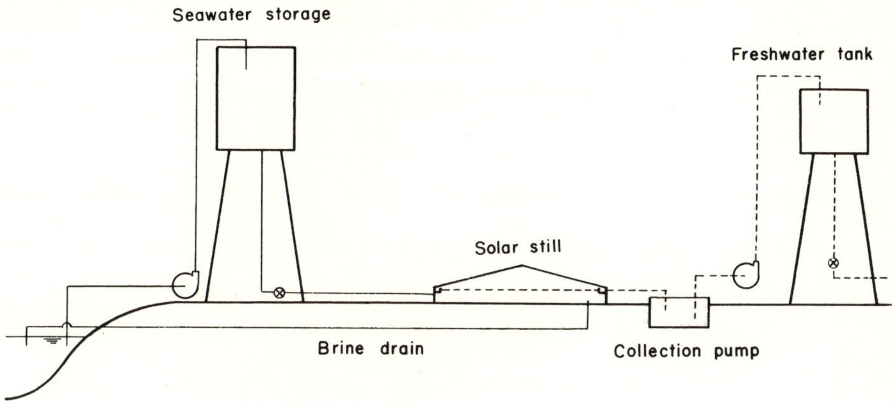

FIG. 10-3. General arrangement of solar distillation plant.

A typical arrangement for this is shown diagrammatically in Fig. 10-3. Saline water may be fed to the still continuously or on a "batch" basis. In the continuous feed, the rate of flow is adjusted so that a calculated ratio between the diurnal rates of brine discharge and distillate production are maintained. In the "batch" system a quantity of water, approximately twice the amount of the distillate expected to be produced, is added at one time during a nonproductive period of the solar cycle. Continuous feed has the advantage of not requiring regular attendance by an operator, but does result in the loss of a small amount of heat energy due to continual discharge of warm water during sunny periods. The batch operation eliminates the heat loss but requires either automatic controls or the presence of an operator to fill the system periodically.

2. Prototype Solar Stills

While solar stills have been studied rather extensively in recent years, there are relatively few in actual use, the largest ones being listed in Table 10-1. Since the major cost of distilled water from solar stills is the capital amortization, the various designs reflect the designer's attempts to minimize the first cost. Schematic cross sections of some of the large solar stills are shown in Fig. 10-4. The deep-basin type, Fig. 10-4a, was developed at Daytona Beach by the OSW from a design by Löf and, in a slightly modified form, was used for the still at Las Marinas, Spain. This design has a continuous basin, lined with asphalt, reinforced concrete beams and pillars for

TABLE 10-1

Large Solar Stills

Country	Location	Basin size (ft^2)	Average productivity (gpd)
Australia	Muresk (II)	4,000	220
	Coober Pedy	34,000	1,680
	Caiguna	4,000	205
	Hamelin Pool	6,000	320[a]
	Griffith	4,450	240
Cape Verde Islands	Santa Maria do Sal	8,000	560[a]
Greece	Aegina	16,040	1,120[a]
	Patmos	93,000	6,900
	Kimolos	27,000	2,000
	Nisiros	22,000	1,600
India	Bhavnagar	4,060	240
Spain	Las Marinas	9,350	680
Tunisia	Chakmou	4,730	140
	Mahdia	14,000	1,100
West Indies	Petit St. Vincent	18,400	1,300

[a]Estimated productivity.

glass support, sheet metal gutters for distillate collection, and is sized to make use of the largest size of commercial glass panes. Figure 10-4b shows the still developed by the CSIRO in Australia and used for installations listed in Table 10-1. This design is based on the use of low-cost glass panes, which limit the width of bays to about 4 ft. The designers chose to use separate bays rather than a

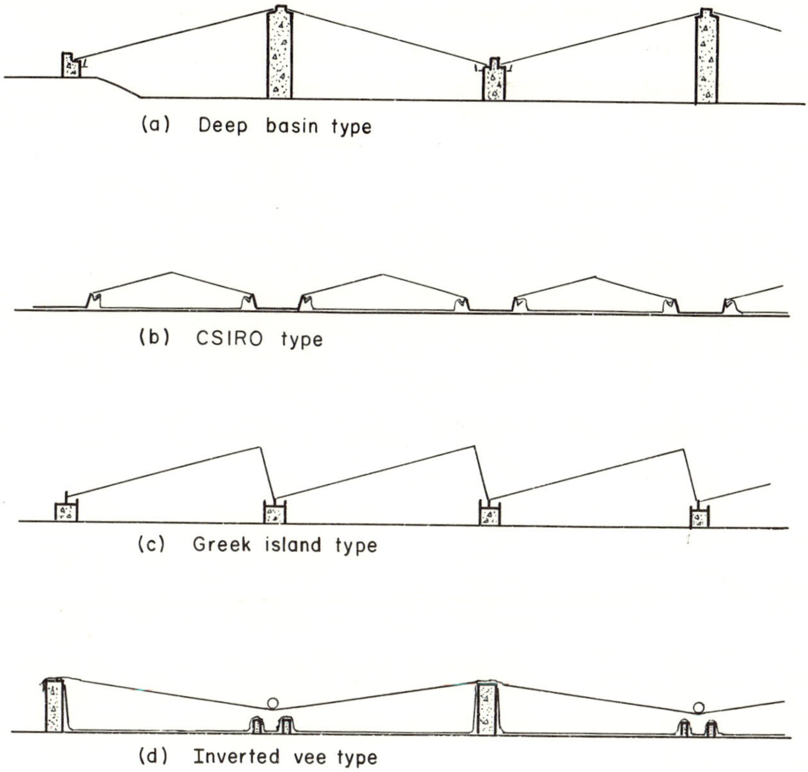

FIG. 10-4. Various configurations of basin-type solar stills. (Schematic only.) Note that (a), (b), and (c) use glass covers; (d) uses sheet plastic cover.

continuous basin. Sheet metal troughs between the bays were designed to support the glass as well as to bound the bays. The upper edges of these troughs were formed to serve as distillate collectors. Each bay was lined with polyethylene to contain the water. Figure 10-4c shows the design developed by Delyannis [13] for use on the islands of Greece. This glass-covered design has separate bays rather than a continuous basin, with concrete foundations supporting the aluminum frames used for securing the glass in position. Figure 10-4d shows the plastic-covered still used on the island of Aegina, Greece. This design also has separate bays, bounded by concrete foundations over which sheet plastic is laid. Longitudinal bars along the center of each bay are sufficiently heavy to depress the plastic sheet and cause the condensate to drip into the distillate channel, which runs along the center of

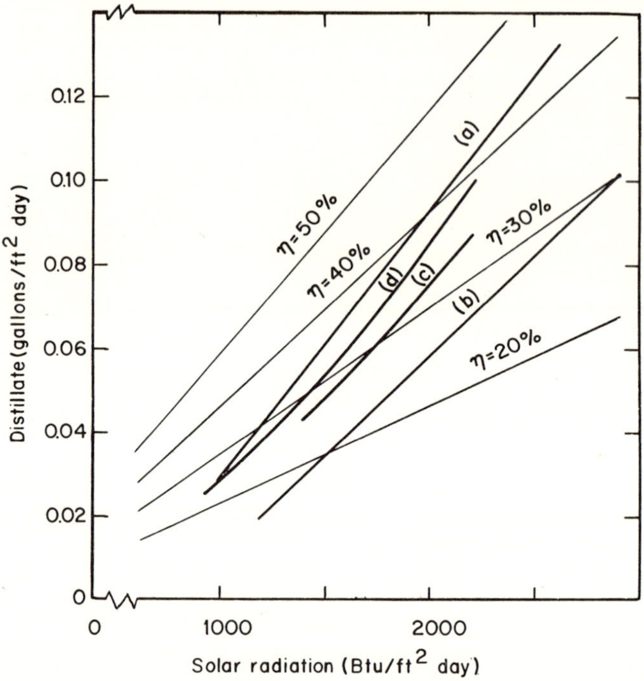

FIG. 10-5. Performance curves for solar stills. (a) Small asbestos-cement still, Richmond, California; (b) sawtooth type still, Richmond, California; (c) deep basin stills, Daytona Beach, Florida; (d) inflated plastic-covered still, Petit St. Vincent Island.

each bay. Although there are numerous other possible variations [14] of solar still configurations, the above examples serve to show the range of such variations.

Sizes of solar stills are given ordinarily in square feet of glass area rather than gpd as for fuel-operated distilling devices. This is due to the highly variable daily output of the solar still. Basin-type solar stills should produce at the rate of about 1 gpd for each 10 ft^2 of glass surface area on sunny days in summer. Because of solar energy fluctuations, yearly averages generally amount to approximately 25 gal/ft^2 for stills in good condition. Hence, if one assumes storage of distilled water sufficient to overcome seasonal variations of daily output, the average capacities in gpd may be determined, as in the third column of Table 10-1.

The performances of many solar stills have been published and plotted to show daily output versus the intensity of solar radiation. From Fig. 10-5 it can be seen that the clear-day points for a representative solar still lie along a sloping straight line, intersecting the x axis to the right of the origin; i.e., zero production rate corresponds to a + value of radiation. Points for days other than clear, i.e., cloudy days, generally lie below the clear-day curve. If a still efficiency is defined as the ratio of the output times the latent heat of the distillate to the incident solar radiation, the radial line on Fig. 10-5 would represent the constant efficiency ratio at various radiation intensities. As indicated there, these lines intersect the clear-day curve at points which show that the efficiency of a solar still increases as the radiation intensity increases.

It must be noted that the performance ratio (PR), previously defined for distillation plants, is very low for solar stills. For example, an output of 0.14 gpd/ft^2 at a solar intensity of 2500 Btu/ft^2 will have a PR of 0.47 lb of water per 1000 Btu, and is to be compared with values of 3 to 10 for fuel-operated plants. This low PR has spurred research into various other schemes for increasing the effectiveness of solar energy usage.

3. Prediction of Performance

The production rate of a solar still for clear-day conditions can be predicted using solar radiation data, the conditions of ambient air, and empirical data for properties of the still components. Equations relating these variables have been proposed by various writers, the report by Talbert et al. [3] containing a summary of these analyses. The scheme of calculation presented in this chapter was originated by Dunkle [4] and modified by Morse and Read [5]. Dunkle proposed seven equations expressing the heat and mass transfer relationships within and around the solar still. Morse and Read showed that these equations could be represented graphically, with a single chart serving to display the ranges of variables commonly encountered. Figure 10-6a shows a cross section of a solar still with arrows indicating the directions of the several energy fluxes. Figure 10-6b is a Sankey diagram showing the manner in which heat flux from the sun is absorbed and dissipated from the solar still. The energy input is the sum of the solar radiation absorbed by the water in the basin, $\alpha_w \tau H_s$, and that absorbed by the transparent cover, $\alpha_g H_s$. The energy reflected by the cover, $(1 - \alpha_g - \tau)H_s$, does not enter into the calculations. The total energy input is partially utilized in increasing the temperature of the system, $c_{wg}(dT_w/dt)$, and the balance of the energy must

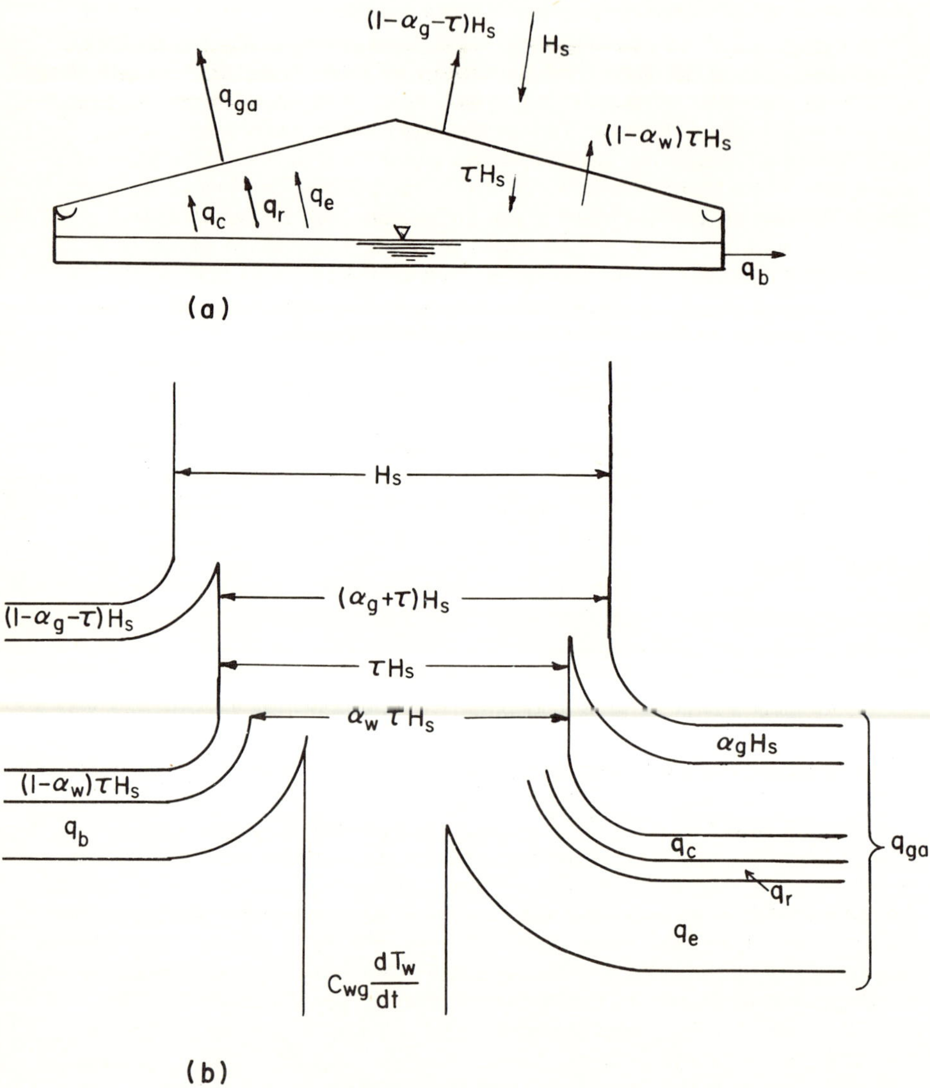

FIG. 10-6. Heat balance diagrams for solar still. (a) Cross section showing direction of heat flux. (b) Sankey diagram for 1 hr during period of water temperature increase.

be dissipated from the cover, q_{ga}, and from the base, q_b. The quantity q_b represents a loss from the system and should be minimized, but q_{ga} must be large enough to discharge the latent heat of condensation of the distillate. It should be noted that the term c_{wg} includes the heat capacity of the water of the still structure and of the ground under the still. The equations as presented by Morse and Read [5] are as follows:

Overall energy balance

$$\alpha_g H_s + \alpha_w \tau H_s = q_{ga} + q_b + c_{wg}(dT_w/dt) \tag{1}$$

Heat balance on transparent cover (internal)

$$q_{ga} = q_r + q_c + q_e + \alpha_g H_s \tag{2}$$

Heat loss from base and edges of still

$$q_b = K_b(T_w - T_b) \tag{3}$$

Heat transfer from water to glass by radiation

$$q_r = 0.9\sigma[(T'_w + 460)^4 - (T_g + 460)^4] \tag{4}$$

σ = Stefan-Boltzmann constant.

Heat transfer from water to glass by convection

$$q_c = 0.128\left[(T_w - T_g) + \left(\frac{p_w - p_{wg}}{39 - p_w}\right)(T_w + 460)\right]^{\frac{1}{3}}(T_w - T_g) \tag{5}$$

Heat transfer from water to cover by evaporation and condensation

$$q_e = 0.0254\left[(T_w - T_g) + \left(\frac{p_w - p_{wg}}{39 - p_w}\right)(T_w + 460)\right]^{\frac{1}{3}}(p_w - p_{wg})h_w \tag{6}$$

Heat balance on transparent cover (external)

$$q_{ga} = 0.9\sigma[(T_g + 460)^4 - (T_a + 460)^4] + h_{ga}(T_g - T_a) \tag{7}$$

By comparing Eq. (1) with the Sankey diagram of Fig. 10-6b it will be observed that the energy reflected from the cover, $(1 - \alpha_g - \tau)H_s$, is

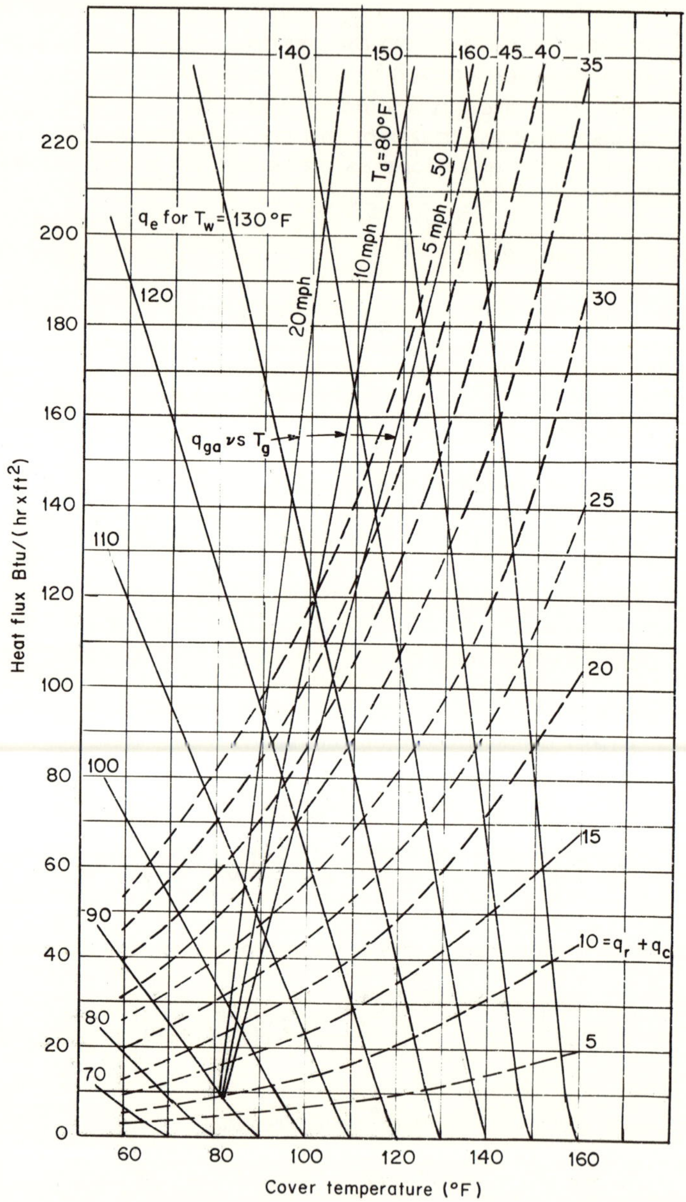

FIG. 10-7. Chart for determining heat flux quantities. Adapted from Morse and Read [5].

excluded. Also the energy reflected from the water surface, $(1 - \alpha_w)\tau H_s$, is assumed to be transmitted outward through the transparent cover, without further absorption or reflection.

The forms of these equations are such that direct solution is impossible, so that it is necessary to employ trial and error methods to solve them. Figure 10-7 utilizes the graph introduced by Morse and Read to reduce the labor of the trial and error process. This figure is used to determine values of q_{ga} and T_g which satisfy Eqs. (2)-(7). The loss through the base, q_b, is computed separately. These values are then inserted in Eq. (1) to obtain hourly changes in T_w. Hourly values of q_e, as determined from Fig. 10-7, are totaled for the 24-hr period to obtain the amount of distilled water produced.

To illustrate the use of Fig. 10-7, the output of a solar still is computed in Table 10-2.

Nomenclature.

H_s	hourly solar radiation values
T_w	temperature of water
T_g	temperature of glass
α_g	absorptivity of glass
τ	transmissivity of glass
α_w	absorptivity of water
q_c	heat transferred by convection between water surface and glass enclosure
q_r	heat transferred by radiation between water surface and glass enclosure
q_e	heat transferred from water surface to glass enclosure by condensation of vapor on the glass
q_{ga}	heat dissipated from external surface of glass enclosure
q_b	heat dissipated through the bottom of the still and from the structure supporting the glass cover

Relations between these quantities are expressed mathematically by Eqs. (1) - (7) and are shown graphically in Fig. 10-6.

TABLE 10-2

Estimation of Solar Still Production for August 15
Rangiroa Atoll, Tuamotu Islands — Latitude 15 °S, Longitude 144 °W

1	2	3	4	5	6	7	8	9	10	11	12	13	14
Time	H_s	T_w	$\alpha_g H_s$	$q_r + q_c$	q_e	q_{ga} (4 + 5 + 6)	q_b	$(1 - \alpha_w)H_s$	(7 + 8 + 9)	$(\alpha_g + \tau)H_s$	$C_{wg}\frac{\Delta T_w}{\Delta t}$	ΔT_w	T_g
07	74	85.0	7.4	2.0	1.5	10.9	5.0	5.9	21.8	66.6	44.8	2.8	82.7
08	146	87.8	14.6	2.8	2.3	19.7	7.8	11.7	39.2	131.4	92.2	5.8	83.2
09	214	93.6	21.4	6.6	5.7	33.7	13.6	17.1	64.4	192.6	128.2	8.1	88.5
10	260	101.7	26.0	13.5	13.7	53.2	21.7	20.5	95.4	234.0	138.6	8.7	93.2
11	296	110.4	29.6	17.5	26.0	73.1	29.2	23.7	126.0	266.4	140.4	8.7	98.0
12	296	117.9	29.6	23.0	42.0	94.6	37.9	23.7	156.2	266.4	110.2	6.9	104.0
13	260	124.8	26.0	28.0	60.0	114.0	44.8	20.8	179.6	234.0	54.4	3.4	108.0
14	214	128.2	21.4	31.0	70.0	122.4	48.2	17.1	187.7	192.6	4.9	0.3	110.0
15	146	128.5	14.6	32.0	74.0	120.6	48.5	11.7	180.8	131.4	-49.4	-3.1	109.0
16	74	125.4	7.4	32.0	68.0	107.4	45.4	5.9	158.7	66.6	-92.1	-5.8	106.0
17		119.6		29.5	57.5	87.0	39.6		126.6		-126.6	-7.9	101.0
18		112.2		24.5	39.0	63.5	32.2		95.7		-95.7	-6.0	96.0
19		106.2		19.5	28.5	48.0	26.2		74.2		-74.2	-4.6	92.0
20		101.6		17.5	23.0	40.5	21.6		62.1		-62.1	-3.9	89.0
21		97.7		14.0	16.0	30.0	17.7		47.7		-47.7	-3.0	87.0
22		94.7		12.0	12.0	24.0	14.7		38.7		-38.7	-2.4	86.0
23		92.3		10.0	8.0	18.0	12.3		30.3		-30.3	-1.9	85.0
24		90.4		8.0	7.2	15.2	10.4		25.6		-25.6	-1.6	84.0
01		88.8		7.5	5.5	13.0	8.8		21.8		-21.8	-1.4	83.2
02		87.4		6.5	4.5	11.0	7.4		18.4		-18.4	-1.2	82.7
03		86.2		5.0	3.9	8.9	6.2		15.1		-15.1	-1.0	82.0
04		85.2		3.5	3.0	6.5	5.2		11.7		-11.7	-0.7	81.8
05		84.5		3.0	2.5	5.5	4.5		10.0		-10.0	-0.6	81.3
06		83.9		2.0	1.8	3.8	3.9		7.7		-7.7	-0.5	80.9
Σ	1980				575.6		512.8						

Constants used in Table 10-2 are as follows:

C_{wg} = 16 Btu/(°F × ft²); i.e., equivalent depth of water

h_w = 1025 Btu/lb = latent heat at T_w average of 120°F

K_b = 1.0 Btu/(hr × ft² × °F)

T_a = 80°F (ambient temperature)

α_g = 0.1 (absorptivity of glass cover)

τ = 0.8 (transmissivity of glass cover)

α_w = 0.9 (absorptivity of water)

T_w = 85°F at 07 hours, windspeed = 5 mph

Using summations from the table the following can be computed:

$$PR = \Sigma q_e/h_w = 575.6/1025 = 0.56 \text{ lb/(ft}^2 \times \text{da); or } 0.067 \text{ gpd/ft}^2$$

$$\text{Efficiency, } \eta = \Sigma q_e/\Sigma H_s = 575.6/1980 = 29\%$$

Values inserted in Table 10-2 were obtained as follows:

1. The value for total daily incoming radiation, ΣH_s, was obtained from Fig. 10-1, and divided into hourly values by the use of the charts of Ref. [2]. These hourly values were recorded in Column 2.

2. At an estimated value of T_g, the intercept $q_{ga} - q_e$ on Fig. 10-7 was compared to the sum $\alpha_g H_s + (q_r + q_c)$, the latter term being also obtained from Fig. 10-7. If these were unequal, other values of T_g were tried until an equality was reached, thus satisfying the conditions expressed in Eq. (2). The resulting values could then be entered in Columns 4, 5, 6, 7, and 14.

3. Terms for the overall energy balance, Eq. (1), were entered in Columns 8 to 12, and the resulting hourly changes in T_w in Column 13.

4. Multiple Use of Solar Energy

The solar stills discussed in the earlier sections of this chapter make use of the solar energy on a single-stage basis, i.e., there is no reuse of the heat energy as in the multiple-effect or multiple-stage flash distillers dealt with in previous chapters. The success of these multiple schemes for increasing the yield of water for a given amount of heat energy in fuel-operated plants has led to numerous attempts to apply similar schemes to plants operated by solar energy. Experimental work in this area has not progressed beyond the laboratory stage to date, for, so far, it has been found that the cost per unit of water produced by the more elaborate plants has not been any less than that of water produced by the simple basin-type still. As already indicated, these costs are of the order of \$3 per kgal from solar stills having a capital cost of about \$30 per daily gallon of capacity. These high costs are acceptable for units of very small capacities, and are less than that of very small fuel-operated distillers of comparable size. However, the unit costs of solar stills do not decrease appreciably as the size is increased, so that large basin-type solar stills are relatively more costly than competitive fuel-operated distillers, the cost of which may be as little as \$1 per daily gallon of capacity. If solar distillers are to become economically interesting for any but the very small installations, now typical of this form, there will have to be a significant decrease in the capital cost, and multiple use of solar energy may be one possible direction of proceeding to accomplish such a reduction, by increasing the yield of water per unit of solar energy collected.

a. Application of Multiple Effect to Solar Stills. The most obvious scheme for increasing efficiency would seem to be the use of the multiple-effect principle. Attempts to apply this principle to solar distillers have resorted to a humidification process rather than one involving deaerated water as is the requirement in conventional multiple-effect and multiple-stage flash installations. Many configurations of design have been devised, and, although they differ in detail, they all contain the basic elements contained in the schematic drawing of Fig. 10-8. A group of parallel metallic plates is arranged vertically behind a transparent window. Saline water is caused to flow down the rear surface of each plate, either in a free-falling film or through some porous material. The space between each pair of plates is filled with air, which is free to circulate. Solar energy, passing through the transparent window, is collected by the blackened surface of the front plate. The resulting heat is conducted through the plate and causes evaporation of part of the water flowing down the rear surface of the plate. The vapor formed is transported as humidity of the air and condensed on the front of the next plate. The heat of condensation is conducted through that plate and produces vapor from the water film on the rear of its surface. The process is repeated as many times as there are plates. The condensate is collected by a gutter along the bottom of each plate and conducted out of the still.

It will be noted that the presence of air in the several compartments between plates makes possible the variation of partial vapor

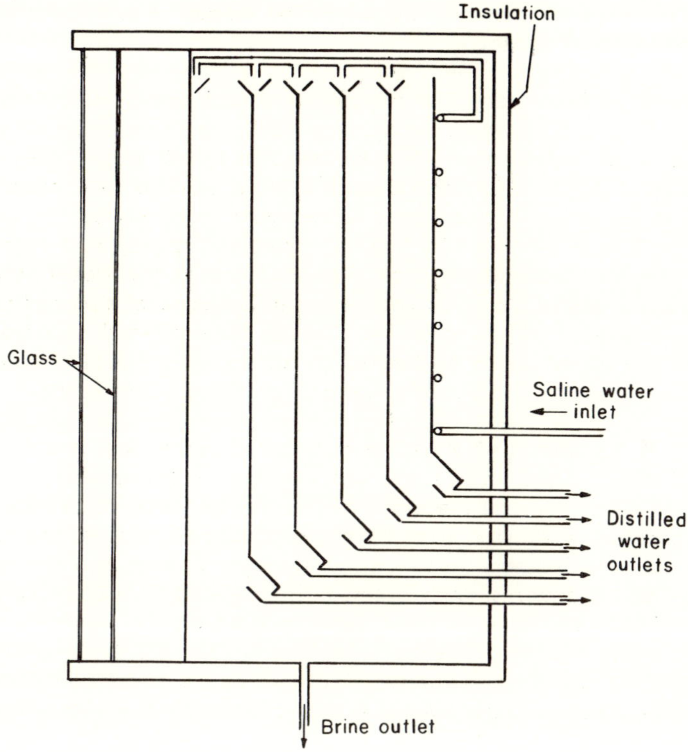

FIG. 10-8. Multiple-effect diffusion solar still adapted from Dunkle [4].

pressure from one "effect" to the next without change in the total pressure. Although this is advantageous from the viewpoint of equipment design, it has the disadvantage that the circulation of the dry air component and the uncondensed humidity creates a wasteful transport of heat between the evaporating and condensing surfaces.

b. Application of Solar Heat Collectors. A second possible scheme for making multiple use of solar energy is the solar heat collector. This device provides a means of heating water to a high temperature by solar energy under conditions in which no evaporation occurs, and the water so heated can then be used in a conventional multiple-effect or multiple-stage flash distiller of essentially the same type as is used with fuel-generated heat. To accomplish this operation the distinguishing feature of such a process would be the solar heat collector, which is entirely separate and independent of the distillation process. The only decision as to whether or not to use this method of heating the water depends upon the relative cost of heating water by means of a solar collector or by conventional fuels.

There have been very few instances where solar collectors have been used for heating water on a large scale, so that most of the available information relating to such an installation is based only on design studies. Representative of such design studies is a report by Brice [6] in which he conceived saline water to be heated in 1000 channels, 48 ft wide and 500 ft long, covering nearly 600 acres of ground. The channels were to be grouped so that the water would flow through several units in series for a distance of approximately 2 miles in a period of about 36 hr. During flow through this path he predicted the water could be heated to as much as 140°F, under conditions that exist in the vicinity near San Diego, California.

Each of the above 48 × 500 ft channels, or basins, was to be a solar collector, lined with some material to absorb radiation, and covered with transparent material which would serve to reduce the low-temperature radiation loss from the absorber. The sides of the basins would be defined by low concrete curbs, with the ground between the curbs covered with black polyethylene film 10 mils thick. This film would serve as a basin liner to contain the water as well as to absorb solar energy. The material chosen for the transparent covers was Tedlar plastic film, selected because of its supposed durability and relative freedom from deterioration caused by the ultraviolet portion of solar radiation. One layer of Tedlar was to rest on the surface of the water to prevent evaporation, and above it were to be one or more layers of the Tedlar film, each layer supported above the one below by a slight air pressure, maintained by air pumps.

In the study cited, Brice considered the effect of multiple layers of transparent material on the amount of solar heat collected. When more than one transparent cover is imposed between the sun and the collecting surface, the surface temperature of the outer transparent cover is lower than that of the inner cover as well as any of the intermediate covers. The more covers there are, the lower will be the temperature of the outer cover, and hence the lower the energy loss therefrom. Partially offsetting this saving in heat energy is the reduction of the amount of solar energy transmitted to the collecting surface, due to absorption and reflection by each of the transparent covers.

The net effect of the use of several transparent covers has been studied and evaluated by many writers, a widely cited article being that by Hottel and Woertz [7]. Using the formulations of this article, Brice [6] calculated the amount of solar energy that would be collected when varying numbers of Tedlar covers were used, and converted these amounts to dollar values. His estimates are based on a useful life for the plastic film of 8 years, 48 years for the concrete curbs, 4% interest rate for amortization, and a period of operation of 8 months from March 1 to October 31 each year. The results of his calculations show that the brine temperature and heat cost in any given month will increase as the number of covers is increased. He concluded that for all the months of operation, the cost of collected heat in a solar collector using five covers would not be more than $0.50 per million Btu. If a distillation plant could be designed to accept a range of brine temperature from about 104°F in October to about 160°F in July and operate efficiently, the use of solar heat could be competitive with fuel heat which costs approximately the same.

The one experimental plant constructed using nonevaporative solar heat collectors was built at Puerta Penasco, Mexico, by Hodges et al. [8]. In this plant five parallel channels were used for solar heat collection, three channels being 6 ft wide and the other two, 8 ft wide. With a length of 300 ft, the total solar collection area was 10,400 ft^2. Black butyl rubber and black polyethylene were both tried as basin liners. A plastic film of PVF was floated on the surface of the water to prevent evaporation, and a second film of Tedlar was fastened along the edges of the channels and supported above the floating film by air pressure. Seawater, pumped into the channel, flowed through at a depth of 2 in. The capital cost of the solar collectors at Puerta Penasco ranged between $0.21 and $0.59 per ft^2, depending on the combination of the films used. The average efficiency of heat collection was about 35%. Assuming that the useful life of the plastic film is 2 years, and the solar rating at San Diego is 600,000 Btu/($ft^2 \times$ year), the cost of capital amortization of the solar collectors plus film replacement would lie between $0.0484 and $0.239 per ft^2 per year, depending upon the film cost noted above. This would be equivalent to $0.23 to $1.14 per million Btu. Cost of pumping would be assigned to the distiller section. As compared with the cost of heat energy from fuel sources, the use of solar collectors of this type is worth consideration where solar radiation conditions are favorable and land costs are reasonable.

C. Use of Ocean Temperature Differences

Since the oceans cover a large fraction of the earth's surface, and water is an excellent absorber of solar energy, it has been argued that a means should be found for utilizing the solar heat which has been absorbed by it. So far, the only schemes advanced for accomplishing this have been based on the observed differences between water temperatures at the surface of the ocean and at considerable depth. Surface water temperatures vary with geographical location and with the season. Temperatures in excess of 80 °F have been observed in a few locations, such as in parts of the Caribbean Sea and along the Ivory Coast of Africa. These same locations have also shown water temperatures as low as 40 °F at depths of 1200 to 1500 ft, thus presenting a potential temperature differential of as much as 40 °F.

Claude [9], as early as 1927, proposed that these temperature differences should be utilized for power generation. His scheme involved the flashing of warm water to produce steam, the expansion of that steam in a turbine to a very low pressure, and finally its condensation at low pressure by the use of cooling water from the ocean's depths. To demonstrate his proposal, Claude installed a small steam turbine on a beach, supplied it with steam flashed from surface water, and condensed the steam by using cold water drawn through a pipe from considerable depth. The turbine was small, developing only about 12 kW, and the demonstration as a power-generating device was terminated after a short period when the cold water pipeline was destroyed by underwater marine disturbances. Nevertheless, this experiment served to demonstrate that both power and distilled water could be obtained from the ocean, but it has not as yet been followed up by any prototype plant.

In the 1940s, the French proposed [10] the construction of a plant using Claude's scheme. The plant was to be located at Abidjan, on the Ivory Coast of West Africa, and was to generate 5000 kW of salable electrical power. Techniques for laying the undersea pipeline were demonstrated, and experiments on the turbines were conducted. However, a competitive scheme for supplying the power needs from a conventional hydroelectric plant in nearby mountains was adopted instead, and the ocean power scheme was abandoned.

Although it was realized that the condensate from a power plant of this type would be pure water, in order to make such water readily available for potable use, the condensers would need to be of the surface rather than of the jet type as employed by Claude. In addition, it became obvious that many chemical and other processing plants along the seacoasts were using ocean water for cooling, and discharging large quantities of waste water at temperatures of 90 ° to 120 °F. Temperatures of this order would permit the use of 60 °F cooling water in the condensers, thereby greatly lessening the problem of installing and maintaining the deep ocean cold water pipeline.

Studies of the problems to be solved in such a plant to be devoted primarily to water production, rather than to power generation, were undertaken at the University of California. The concept adopted for these studies was to generate steam by flashing, to expand the steam in a turbine capable of producing only enough power to run the plant machinery, and to liquefy the steam in a surface condenser. The chief problems for which solutions were sought were (1) to obtain vapor release with a minimum of salt fog production, (2) to eliminate the dissolved air from the steam before the condensation process, and (3) to study the use of mechanical vacuum pumps for air removal rather than jet pumps. Solutions to these problems were published [11] at the conclusion of the experimental work. It was found that the two chief drawbacks to the application of this scheme for water production from waste cooling water seemed to be the comparatively small capacities that are possible with water production facilities of this type, and the extremely high vacuum which must be maintained in the condensers. So far no plants of this type have been built, in spite of the attempts by at least two manufacturers to place them on the market.

A related process, in which electric power is generated by means of a unique application of the temperature differential of ocean water, has been suggested by Anderson [12]. His proposal was to use the hydrocarbon, propane, rather than water, as the active working fluid. Warm water from the ocean surface is circulated through a "boiler" in which the propane vapor is produced. This vapor expands through a power-generating turbine and is passed on to be condensed in a surface condenser, cooled by water from the deep ocean. The power generated is a cheap source of electrical energy which can be used to run a combined vapor compression distillation plant.

A novel detail of his proposal involves the use of a floating power station, constructed in somewhat similar fashion to the craft used for offshore oil-well drilling. The large hull, housing the power turbines and providing living quarters for the crew, would also serve to collect the surface water and funnel it into the vertical pipes leading to the propane "boilers." Both the "boilers" and the condensers would be suspended below the hull at a depth such that the hydrostatic head was equal to the propane vapor pressure in the "boilers" and the propane vapor pressure inside the condensers. The cold water pipe would extend vertically below the condenser to a depth sufficiently great to obtain water at a temperature of about 43°F. Anderson's estimates for a 100,000 kW plant of this type showed a cost of $166 per kW, corresponding to a power cost of a little less than $0.003 per kWh. Although no plant of this type has as yet been constructed, it offers a challenge for both researchers and manufacturers.

II. UTILIZATION OF GEOTHERMAL ENERGY

Besides solar energy there is available some other terrestrial energy in specific and scattered locations. Geologists, involved in studies of the structure of the earth, have hypothesized that tremendous amounts of heat energy are generated by radioactive decay in the liquid outer core of the earth. This heat is conveyed through the surrounding mantle by conduction and possible convection of highly viscous or plastic material, and delivered to the basaltic rock which constitutes the bottom of the earth's crust. The heat is then conducted through the rock and the overlying sediments or oceans, to escape into the atmosphere. Temperature gradients observed in mines and oil wells, reflecting this hypothesis, show an average of about 1 °F per 100 ft of depth in most of the earth's surface. In a few locations, gradients of as much as 12° to 25 °F per 100 ft have been found, generally in areas near hot springs, volcanos, and earthquake faults. These areas of large gradients, with their potentials of hot water and steam, offer challenging sources of heat energy for the production of power and freshwater.

The use of geothermal heat to date has found some application to power generation and space heating. Substantial power generation installations are located at Lardarello, Italy, at Wairaki, New Zealand, and at The Geysers in northern California. Hot water and steam for space heating is widely used in Iceland, and to a much more limited extent in isolated locations such as Rotorua, New Zealand, Calistoga, California, and Klamath Falls, Oregon.

REFERENCES

1. H.G. Houghton, On the annual heat balance of the northern hemisphere, J. Meteorol., 11, No. 1, Feb., 1954.
2. A. Whillier, Solar Radiation Graphs, Solar Energy, 9 (No. 3), 1965, pp. 164-165.
3. S.G. Talbert, J.H. Eibling, and G.O.G. Löf, Manual on Solar Distillation of Saline Water, Res. Develop. Progr. Rept. No. 546, Office of Saline Water, U.S. Dept. Interior, 1970.
4. R.V. Dunkle, Solar Water Distillation: The Roof Type Still and a Multiple-effect Diffusion Still, Intern. Develop. Heat Transfer, 1961, pp. 895-902.

5. R. N. Morse and W. R. W. Read, The Development of a Solar Still for Australian Conditions, Mech. Chem. Eng. Trans., Inst. of Engineers, Australia, May, 1967, pp. 71-80.
6. D. B. Brice, Saline Water Demineralization by Flash Evaporation Utilizing Solar Energy, Advan. Chem. Ser. 38, 1963, pp. 99-116.
7. H. C. Hottel and B. B. Woertz, Trans. ASME, 64, 1942, p. 91.
8. C. N. Hodges et al., Solar Distillation Utilizing Multiple-Effect Humidification, Res. Develop. Progr. Rept., No. 194, Office of Saline Water, U. S. Dept. Interior, May, 1966.
9. G. Claude, Power from the tropical seas, Mech. Engineering, 52, No. 12, 1930, p. 1039.
10. M. A. Nizery, Etude sur les possibilities d'utilisation de l'energie thermique des mers et de l'energie solaire, Bulletin de l'Institut Oceanographique, No. 906, 30 Dec. 1946, Monaco.
11. E. I. Ewoldsen et al., Sea water entrainment in low temperature flash evaporators, Report no. 64-3, Sea Water Conversion Laboratory, Univ. of California, Berkeley, Dec., 1964.
12. J. H. Anderson, The sea plant – a source of power, water and food without pollution, Solar Energy, 14, No. 3, 1973, pp. 287-300.
13. A. Delyannis and E. Piperoglou, The Patmos Distillation Plant, Solar Energy, 12, No. 1, 1968, pp. 113-115.
14. M. Telkes, Solar Stills, Proceedings of World Symposium on Applied Solar Energy, Stanford Research Institute, 1956, pp. 73-81.

Chapter 11

FREEZE AND HYDRATE SEPARATION

I. STATUS OF FREEZE-SEPARATION PROCESSES

As indicated in Chapter 4, desalination by freeze separation is possible because the ice crystals formed by removal of heat from saline water exclude the ions of dissolved salts. Freeze separation is attractive because the low temperatures involved greatly reduce the problems of scale formation and corrosion, which are so troublesome in all forms of distillation. This attractive feature of the process, together with the prospect of low energy requirements as compared with distillation, has stimulated considerable investigation. A related process, known as hydrate-separation, has similar advantages, and the use of materials that can be applied to hydrate separation has been the subject of much research. Crystalline hydrates of materials, such as propane, form at temperatures somewhat higher than the freezing point of water, and thereby reduce the problem of precooling the saline water, as compared with using water or other refrigerants for producing ice. In spite of these advantages, however, neither freeze separation nor hydrate separation has been developed to date to the point that they are being widely used. The one freeze-separation plant installed for water supply, located in Eilath, Israel [1], is designed with four units and has a capacity of 250, 000 gpd. After its operation for several years, a nearby 1-mgd distillation plant was built, and the freeze-separation plant was placed in "mothballs" pending the increase of water demand above the 1-mgd level. The manufacturer of this plant,

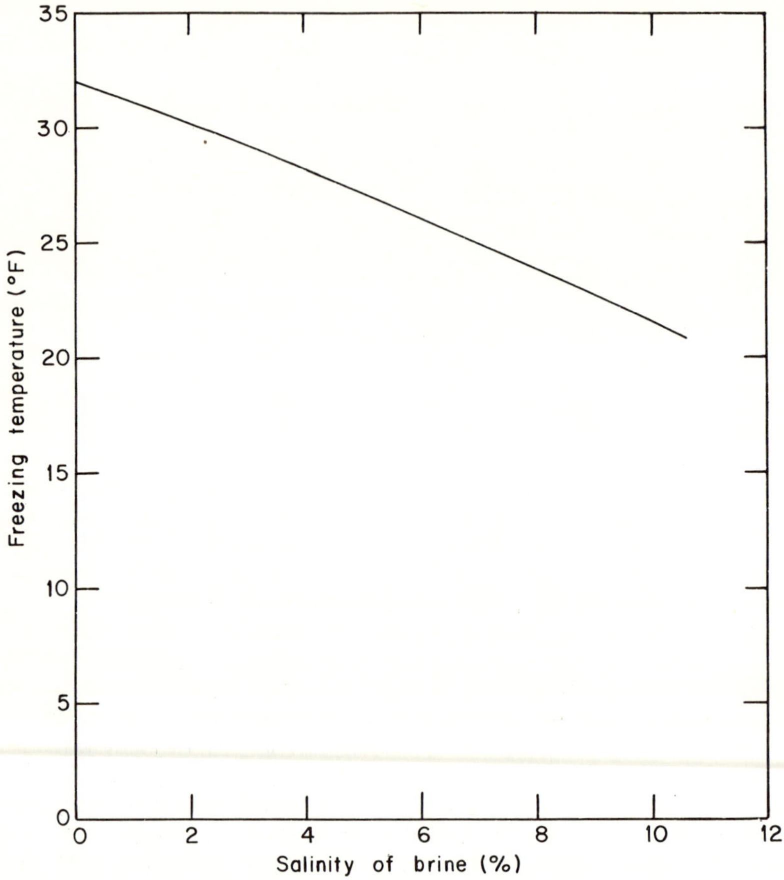

FIG. 11-1. Freezing temperature of seawater and its concentrates.

Colt Industries, has continued to further develop the equipment, and has demonstrated reliable performance of units with 100,000 gpd capacity and has proposed units with 500,000 gpd capacity [2]. During the same period, development work has been done in Israel, as reported by Pachter and Barak [3]. Their published results indicate that compressor efficiencies of 60% have been achieved and that other plant components have been improved to the extent that capital costs can be expected to be as low as those of the MSF distillation plants and that energy requirements are competitive with those of vapor compression distillers. In spite of the developments noted above, there is no apparent trend toward the installation of freeze- or hydrate-separation plants in preference to distillation plants, so that these processes must be regarded as still in the experimental pilot plant stage.

A. Freeze-Separation Components

Freeze-separation processes consist of a heat-removal portion to produce ice crystals, a separatory portion to separate the ice crystals from the brine, and a heat-addition portion in which the ice crystals are melted to give liquid water.

1. Schemes for Ice Production

The temperature at which heat must be removed to form ice in saline water is less than the freezing point of pure water, because the presence of dissolved materials lowers the freezing temperature. The amount of lowering of the freezing point can be observed from the curve for seawater shown on Fig. 11-1, which shows the freezing temperature as a function of the salinity of seawater solutions. For example, the curve indicates that normal seawater, 3.5% salinity, will begin to freeze at 28.5°F and that a brine of twice this concentration, 7% salinity, would begin to freeze at 25°F. From these figures it can be concluded that any freeze-separation scheme for removing 50% of the water from seawater would involve cooling the water to 25°F, and withdrawing heat at this temperature. Since these temperatures are less than the ambient temperature, refrigeration is required to remove heat for the formation of ice crystals. The various ways in which regrigeration can be applied to the cooling and freezing of water has given rise to several schemes for desalination which include (1) the application of some of the machinery developed for the ice-making industry, (2) the use of water itself as a refrigerant, and (3) the use of other refrigerants in direct contact with the water being frozen. The first efforts in desalination by freeze separation lay in the direction of applying conventional ice-making machines to the production of ice from saline water. Moulton and Hendrickson [4] applied several ice-making machines to the freezing of saline water and concluded that they were not suitable for desalination purposes, primarily because of high cost due to the fact that the heat removed had to pass through a solid surface as well as two liquid films.

In an effort to eliminate solid surfaces, proposals were made to use water itself as a refrigerant. Such a process depends on the production of water vapor for the absorption of heat. The very low pressure of this water vapor imposes serious problems of compressor size which, as will be seen later, can be improved by the use of other refrigerants. To gain some appreciation of this problem, reference is made to the phase diagram for water, in which the temperature-pressure relationships are shown on Fig. 11-2. Legends on the diagram indicate the triple point, A, where pure liquid, vapor, and solid are in

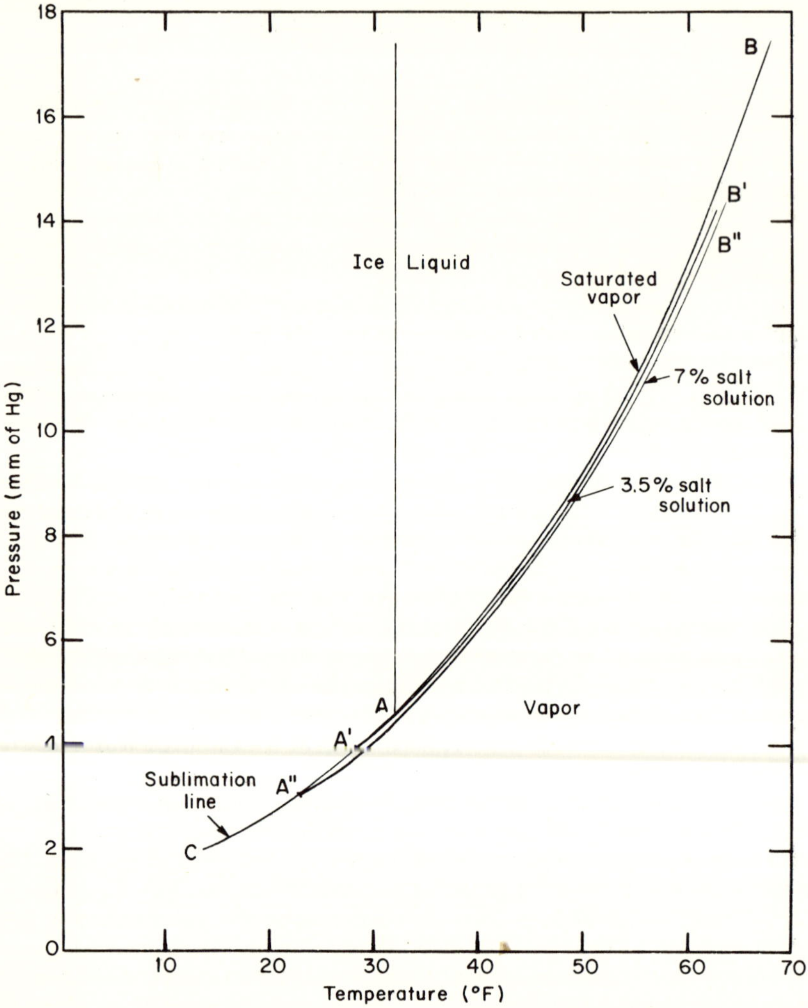

FIG. 11-2. Phase diagram for water with vapor-pressure curves for seawater.

equilibrium. The saturated vapor line, AB, separates the pure water liquid and vapor regions, and the sublimation line, AC, separates the vapor and solid regions. The presence of any solute in the water moves the triple point to A', at a temperature below that of A, but does not move the sublimation curve appreciably. The saturation curves, A'B' and A"B", are drawn for seawater and twice-concentrated seawater.

Intersection of these curves with the sublimation curve, AC, give the approximate temperature limits for the formation of ice crystals under equilibrium conditions.

From this diagram it will be seen that the pressure of water vapor formed when ice crystals are frozen at 25°F is approximately 3.4 mm of mercury, absolute. At this pressure the specific volume of vapor is about 3500 ft^3/lb. This vapor must be removed from the freezing chamber in order for the process to continue, either by passing through a compressor as in the Zarchin process, on which the Colt Industries development is based, or by dissolving in the concentrated solution phase of an absorption-refrigeration cycle, as in the Carrier Corporation experimental plant.

The use of refrigerants other than water in direct contact with the saline water does away with the limitations imposed by the very low density of water vapor noted above, since the specific volume of most refrigerants is far less than that of water. The desalination refrigerant on which most research has been done is n-butane. This material has a vapor pressure of 13.0 psia at 25°F, and the vapor produced occupies 6.6 ft^3/lb at this pressure. With a latent heat of evaporation of 166 Btu/lb, slightly less than 1 lb of n-butane must be evaporated for each pound of ice formed. In comparison, the use of water as a refrigerant, described in the foregoing paragraph, would necessitate the evaporation of only 0.135 lb of water per pound of ice produced, but the vapor would occupy about 470 ft^3. This decrease of about 80-fold in the size of compressor needed has led to the development of several desalination schemes. In all of these the liquid n-butane and water mixture is cooled under pressure, and the mixture discharged through an orifice into a freezing chamber in which the pressure could be maintained slightly below the saturation vapor pressure of the n-butane. The resulting flash evaporation of the n-butane extracts the heat from the water and causes part of it to freeze. The n-butane vapor is then pumped out of the freezing compartment and passed into a condenser for liquefication and recycling.

From Fig. 11-2 we see that there must be cooling below the equilibrium temperatures to produce crystals. There must also be a nucleating agent of some sort. Nuclei may be present in the form of suspended solids, or may result from stirring or other changes in the physical conditions. Crystal growth is relatively slow, so that some residence time must be allowed in the freeze chamber, the amount of time being determined by the geometry of the chamber and the conditions of flow of the several phases.

2. Methods of Separating Ice and Brine

The methods of producing ice, as described in the previous section, result in small crystals mixed in the brine from which they are produced. It should be emphasized that the ice in a desalination process is not in large pieces such as ice cubes, but is more like a very wet snow. The ice crystals formed from seawater appear to be oval or hexagonal plates, from 0.1 to 0.8 mm in length and of a thickness less than 1/5 the length. Their relative dimensions give a very high ratio of surface area to volume, which is quite desirable for the melting portion of the process, but makes it difficult to separate the ice from the brine. When a mass of ice crystals is filtered out of seawater brine, the high surface tension of the water causes the brine to cling to the surface of each crystal as well as trapping some of the brine in the space between crystals. The total weight of brine in a drained mass of such crystals is from 40 to 100% of the weight of the crystals.

Three methods have been used to separate ice from brine in freeze-separation processes, namely, centrifugation, compression, and countercurrent washing. Centrifugation was attempted first and proved to require excessive amounts of energy to operate the equipment, and to be relatively ineffective for producing high purity water. Moulton and Hendrickson [4], in their studies of the compression method, found that the application of pressures from 1000 to 2000 psi could reduce the salinity of melted ice to 500 ppm and yield a quantity of water equivalent to 90% of the ice. Their pressure-type experiments to purify the ice were made using "batch" feed and have not been followed by development of apparatus for "continuous" feed, probably because of the relatively large amount of energy required to accomplish the separation.

The greatest success in separating the ice from the brine has been achieved by the use of countercurrent wash columns, in which a small portion of the product water, flowing in a direction counter to that of the ice motion, is used to wash the ice. The vertical wash columns have been both circular and square in cross section, with pressure-controlled brine outlets located 2 or 3 ft below the top of the columns. In most cases the brine outlets are screened sections in the sides of the columns, although one company has used vertical pipes extending from the bottom and terminating at the desired level in screened openings. In operation, the ice-brine slurry from the freezing chamber is pumped continously into the bottom of the wash column. The ice crystals rise due to buoyancy and form a porous ice pack, which extends from the top of the column to a foot or two below the brine outlet. The brine, in passing through the pores of the ice pack to reach the brine outlet, exerts an upward force on the pack due to fluid friction and thus feeds it toward the ice removal device at the top of the column. About 5% of the product water is used as wash water, washing off the brine from the crystals as it flows downward from the top of the column to join the waste brine at the brine outlet. The device for removal of the ice to the melter has been generally a scraper blade, rotated over the upper washed end of the ice pack, shaving off the clean ice and passing it through a chute into the melting unit.

An alternative ice-removal scheme has been proposed to avoid moving the ice from the wash column to a melting chamber. In this scheme the ice in the wash column is removed by melting rather than by mechanical means. This proposal resulted from the observation that most of the brine leaves the wash column at the level of the brine outlet, and that there is a distinct interface slightly above this level between the rising brine and the down-flowing wash water. Because of this interface it was evident that the brine did not progress further up the column, and that the wash water and ice near the top surface of the pack was relatively pure. If the vapor for melting the ice is brought into contact with the top of the ice pack, the melt resulting therefrom can be drawn off at an appropriate distance from the top as product without interfering with the wash operation. This scheme was proposed by the Colt Industries [2] for use in large plants, based on units of 0.5 mgd each. In their concept, sharp blades would be placed below the top of the column to split the ice pack into a number of small columns, thus permitting the vapor to come into contact with a larger area of ice and also allowing the melt water to drain toward the sides of the column.

3. Methods for Melting the Ice Crystals

All schemes for melting the ice are similar, in that condensing vapor is the source of heat. When water is used as the refrigerant in the freezing compartment, the water vapor which is formed in that chamber is pumped to the melter, where it condenses on the surface of the ice crystals, thus causing the crystals to melt. In this case the condensed vapor adds to the amount of the melted product. However, because the volume of the vapor involved in the case of water as refrigerant is so large, designers have tried to minimize the distance through which it must travel on its way to the melter. An example of such an attempt is the "hydroconverter," proposed by Colt Industries, [5]. In their design, the melting chamber is an annular space above the freezing chamber and surrounds the passage through which the vapor passes in moving to the centrifugal compressor at the top of the vapor passage, and from which the vapor is discharged downward onto the ice in the melting chamber. Thus, in this manner the distance the

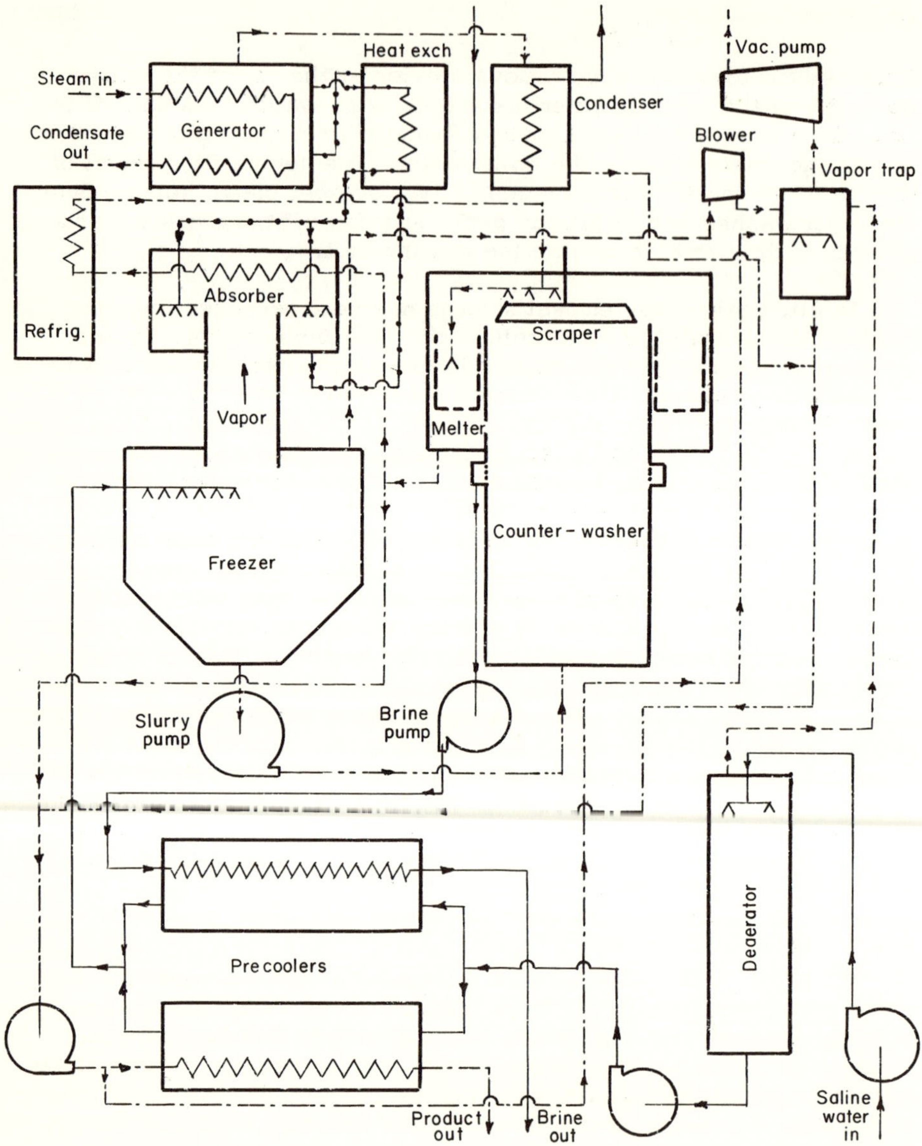

FIG. 11-3. Vacuum flash vapor absorption type of freeze-separation desalting plant (schematic only). ——, Saline water and brine; —• •—, slurry ice and brine; —•—, freshwater and vapor; — — —, air and vapor mix; —•—•—, refrigerant.

vapor travels is limited to roughly only twice the height of the melting compartment.

When a refrigerant other than water is used, such as isobutane, the vapor removed from the freezing chamber is also sent to the melter where it condenses on the ice crystals, causing them to melt. In this case, however, the liquid resulting from the condensation of this vapor does not add to the product, but must be separated from the latter and recirculated to the freezer. The refrigerants used are selected for their lack of solubility in water and the ease by which they can be separated from the product by decantation.

B. Freeze-Separation Plants

The possible schemes for producing ice crystals, separating them from the brine, and melting the purified crystals are numerous, but only three combinations have been significantly developed to date. These are the "vacuum flash vapor absorption system," the "vacuum flash vapor compression system," and the "refrigerant flash vapor compression system."

1. The Vacuum Flash Vapor Absorption System

This type of system utilizes heat transfer by direct contact by means of an absorption refrigeration machine. The only installation of any importance, using this idea, was the experimental plant built and tested by the Carrier Corporation at Syracuse, New York in 1959 [6] and had a capacity of 15,000 gpd. Figure 11-3 is a schematic diagram of such a plant and illustrates the principles involved in its operation. It will be noted that this plant contains the three basic components—a freezing chamber and a combined wash column and melter, together with a lithium bromide absorption system for absorbing the water vapor, a refrigeration system for discharging the energy and heat leakage, and a deaerator and precooler for preparing the feedwater. In operation, the deaerated and precooled saline water is sprayed into the vacuum freezing chamber where part of it vaporizes and thereby cools the remainder, causing ice crystals to form. These ice crystals and residual brine are pumped out of the bottom of the freezing chamber as a slurry and delivered to the wash column. The latter is of the type already described and is equipped with a rotating scraper which discharges the shaved ice to an annular melting chamber which surrounds the upper portion of the wash column. Some of the previously made product water, which has been heated by passage through the condenser of the

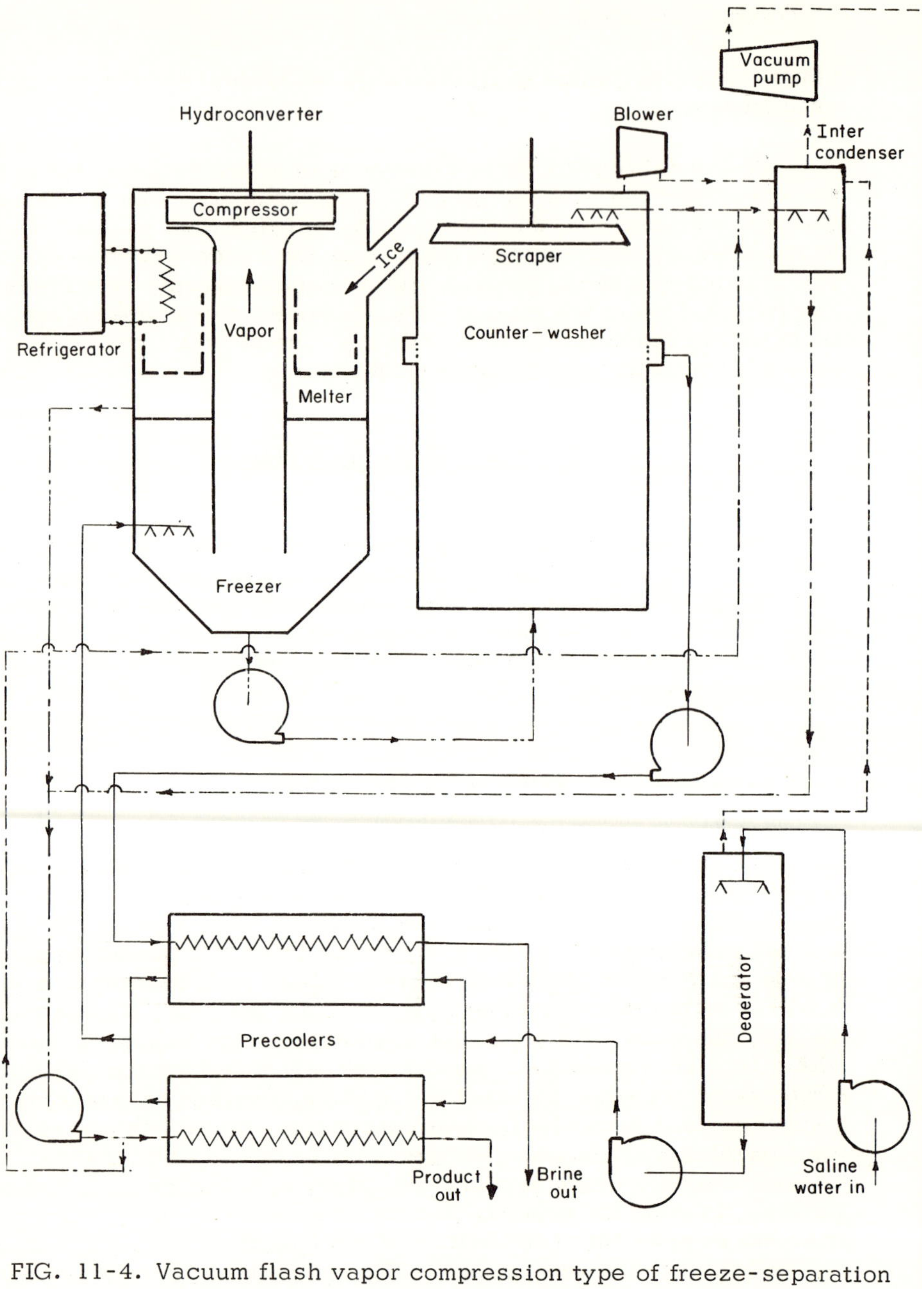

FIG. 11-4. Vacuum flash vapor compression type of freeze-separation desalting plant (schematic only). ———, Saline water and brine; —‧‧—, slurry ice and brine; —‧—, freshwater and vapor; — — —, air and vapor mix; —‧—‧—, refrigerant.

refrigeration system, is sprayed over the ice and causes it to melt. Upon leaving the melting chamber, the mixture of melt water and recirculating pure water is divided, with part of the water going to the precooler as product water and the balance going first to a cooling coil in the absorber and then to the condenser of the refrigerator. The water vapor formed during the freezing process passes into the absorber where it is condensed by absorption in a spray of the concentrated lithium bromide solution. The resulting diluted solution of lithium bromide passes through a heat exchanger and then to the steam-heated generator of the absorption system, where water is again vaporized and delivered to a condenser, cooled by a separate stream of seawater. The condensed vapor is discharged to mix with the outgoing product water from the melting tank, so that the product stream consists of the melted ice plus the vapor formed during the freezing process. The brine from the wash column is discharged through the precooler system to prepare the feedwater.

Although it has been demonstrated that the above plant could produce freshwater, the high capital and operating costs, largely due to the high cost of the absorption refrigeration machine, proved to be so great that no commercial application has resulted to date.

2. The Vacuum Flash Vapor Compression System

The second scheme for freeze separation is similar in several respects to the vacuum flash vapor absorption system already described, but differs from it in that vapor compression rather than vapor absorption for removing vapor from the freezing chamber is used. Invented by A. Zarchin [7] and developed by the Fairbanks Morse Division of Colt Industries [5], the plant arrangement is shown schematically in Fig. 11-4, and is designated by its developers as a VFVC, or vacuum freeze vapor compression plant. This type of plant uses water as the refrigerant and hence several of the components operate under high vacuum conditions. It will be noted that in comparison with the previous system the number of components is smaller in this scheme. The diagram shows two major units, the hydroconverter, which combines the freezing and melting units, and the wash column. In addition, there are auxiliary units for deaeration, for precooling the feedwater, and for discharging heat from the system.

In operation, precooled and deaerated saline water is sprayed into the vacuum freezing chamber in the lower portion of the hydroconverter, the operation of which has already been described. Ice formed in the freezing chamber is pumped as a slurry into the bottom of the counterwasher, also following the purification routine previously outlined.

The rotating scraper blade at the top of the counterwasher discharges the ice into a chute connected with the melting chamber of the hydroconverter, where it comes into contact with the vapor from the compressor discharge. The melt water leaving the hydroconverter passes through the heat exchanger and is increased in temperature as it precools part of the feedwater, the remainder of the feedwater being precooled by exchanging heat with the outgoing waste brine. It will be noted that the hydroconverter and counterwasher must both be kept under vacuum conditions, while in the previously described plant the counterwasher and melter were not exposed to vacuum conditions.

The vacuum pumping system is located on top of the counterwasher. It consists of a blower, an interstage condenser, and a vacuum pump. The blower pumps air and vapor out of the counterwasher, the air representing in-leakage through the entire system. The blower discharges air vapor mixture into a spray chamber in which most of the vapor is condensed by contact with a spray of pure product water. The resulting condensate is added to a portion of the product water and is discharged to the counterwasher for washing the ice as already described. The packed column deaerator is also vented into this spray chamber, so that the vacuum pump connected to this chamber collects the residual air from the incoming feedwater as well as from in-leakage throughout the system.

Two heat exchangers for precooling that are used in most of the existing plants are built of a largo number of thin aluminum plates, stacked parallel to each other and spaced apart by dimples in the plates, the spacing being slightly less than 0.2 in. Alternate flow passages between the stacked plates are manifolded together to form two separate channels in each exchanger—one channel for the incoming seawater and the other for waste brine in one exchanger and for product water in the other. The edges of the plates are sealed by peripheral gaskets with ports on opposite edges for admitting and discharging water. As shown in Fig. 11-4, part of the incoming seawater is cooled by the waste brine in one heat exchanger stack, and the balance of the incoming seawater is cooled by the outgoing product water in the other heat exchanger stack. This flat plate design is regarded as more compact and less expensive than equivalent shell and tube exchangers, although the latter type could have been used for liquid-to-liquid heat exchange operation.

Because of the high vapor volumes involved in the process, namely, 300,000 ft^3/min of vapor for a water production rate of 100,000 gpd, the Colt design of the vapor compressor used in their VFVC plant is a vital cost-saving component. The compressor blades in this plant are of thin, flexible stainless steel sheets, and are supported so that they can take different curved positions for different speeds and flow rates. Hittman Associated [8] records an example of such a vapor compressor in a 100,000-gpd plant having blades approximately 4 ft long and operating at 3000 to 3500 rpm with an adiabatic efficiency of about 60%. The plant built at Eilath was an early version of the VFVC design.

3. Refrigerant Flash Vapor Compression System

The third type of freeze-separation system involves the production of ice crystals by the use of a refrigerant other than water. Its principal advantage over the VFVC process is that it avoids the problems arising from the high vacuum conditions required in the latter process.

The use of refrigerants other than water has been investigated extensively under the auspices of the OSW [9]. Freezing in this process is accomplished by releasing the cold liquid refrigerant into precooled saline water in such a fashion that a maximum surface of water comes into contact with the liquid refrigerant. The scheme which has worked best in the several pilot plants built to test the process has been that of pumping the refrigerant through a submerged nozzle and thereby producing a great deal of turbulence in the precooled feedwater. Ice crystals formed by this means are about 1 mm in greatest dimension and are readily washed because of this large size. The washing process is similar to that of the previously described systems. The melting operation is accomplished by exposing the ice crystals to compressed refrigerant vapor. This was achieved, in one of the pilot plants tested, in a tray-type tower [10], in which ice crystals and recirculated product water were fed to the top of the column as a slurry, and the refrigerant vapor entered near the bottom. The resultant mixture of melt water and liquid refrigerant was removed from the bottom of the tower and sent to a decanter for separation. The excess of refrigerant vapor was taken out of the top of the tower, recompressed, and then liquefied in a condenser before being returned to the freezing chamber. Several refrigerants have been used, the most favored one being n-butane. In spite of the fact that the compressor for immiscible refrigerant vapor may be much smaller than that for the water vapor in the Zarchin process, the one attempt to construct a large plant using this scheme did not prove successful, and development along this line has been relatively unproductive.

C. Energy Requirements for Freeze Separation

As indicated in the opening paragraph of this chapter, freeze separation was thought to be attractive because, basically, it requires less energy than that needed for distillation. This supposition is based on the fact that the latent heat of fusion involved in the formation and melting of ice is only about 1/7 as much as the latent heat of vaporization. However, this advantage does not seem to be borne out in practical application. Distillation schemes are all carried out at temperatures above ambient, so that the energy added to a distillation plant can be more readily dissipated to the surroundings than can that of a freeze-separation plant. In the latter, freeze separation must be accomplished with energy added to the plant at temperatures below the ambient. As a result the energy added to the plant, and which must be dissipated, will need to be pumped from the freezing temperature to the ambient temperature. This involves substantial additional energy for the operation of a heat pump, thereby partially offsetting the anticipated advantage.

To illustrate this situation, the energy required by a freeze-separation process can be compared with that required by a vapor compression process. The vapor compression process can be assumed to involve the distillation of seawater at 212°F in surroundings at a temperature of 70°F. The recovery ratio is assumed to be 50%, and the boiling point rise of the brine during vapor production is 2.1°F. The minimum energy required by this process is that required for isentropic compression of the water vapor to atmospheric pressure, the condensation of which is 212°F. This amount of energy is found from steam tables to be 4.0 Btu/lb or 9.76 kW h/kgal.

For comparison, the freeze-separation plant for seawater at the same recovery ratio must operate at a temperature of 24.5°F in the freezer and 32°F in the melter. Refrigerant vapor formed at 24.5°F during the production of the ice crystals must be compressed to 32°F to melt the ice, and the energy thereby introduced into the system must be delivered to the ambient temperature of 70°F. While the energy required will vary slightly with the refrigerant, a calculation made for any of the refrigerants will give the order of magnitude. For example, Freon-12 is a refrigerant, the property values of which are generally available. The latent heat of Freon-12 at 24.5°F is 67.45 Btu/lb, and the enthalpy and entropy of saturated vapor are, respectively, 80.99 Btu/lb and 0.16921 Btu/(lb × °R). The saturated vapor pressures at 32° and 70°F are respectively, 44.86 and 84.82 psia. Isentropic compression of Freon-12 vapor from 24.5° to 32°F requires about 4.4 Btu/lb, which is approximately the same as that required for the above vapor compression example. Thus it can be seen that freeze separation does not appear to have energy economy superiority over distillation. This is reflected in the fact that as of 1971 very few commercial plants of the freeze-separation type have been built as compared to the large number of vapor compression plants.

D. Zone-Purification Concept

While the generally accepted device for separating ice from the brine is the wash column, there has recently been considerable investigation of an entirely different process, analagous to that used in refining metals, and which is referred to as the "zone-purification" concept. This concept had its origin from observations of Arctic explorers, who found that ice in an iceberg, when melted, was relatively high in purity due to a "weathering" process. It was reasoned that these icebergs, formed from seawater during the intense cold of the Arctic winter, originally contained pure ice crystals surrounded by pockets of brine. Slight differences in temperature between the seawater and air caused these brine pockets in the iceberg to migrate toward the surface of higher temperature in a manner similar to what is known as "zone purification" of metals, with the result that the iceberg gradually became free of the brine.

Laboratory studies of frozen seawater have duplicated this condition. This zone-purification concept can best be explained as follows. Suppose an endless belt of inert material, such as neoprene, is caused to move slowly and horizontally in a refrigerated box cold enough to cause seawater to freeze solid. Narrow depressions extend crosswise of the belt but do not reach either side of it, so that seawater poured into these depressions at the beginning of the belt travel will soon become sticks of ice. As the belt moves along it passes under a number of heating bars placed diagonally with respect to the belt travel. When each stick of ice passes under a heater bar the ice under the heater bar melts and then refreezes, so that a zone of liquid brine moves from the refrozen end of the stick of ice to the warmer end. By repeating the process under successive heater bars, the dissolved salts of the liquid brine, being more soluble in liquid than in crystalline ice, become more concentrated at the ends of the sticks toward which the liquid moves than at the other ends. These salty ends can then be discarded and the balance of the ice melted to give relatively pure water.

The process pictured above can be carried out by many mechanical schemes other than the endless belt used in the explanation. Himes et al. [11] considered a number of possible arrangements and chose the use of vertical finned tubes as the most likely scheme to lead to a practical desalting device. In this arrangement, saline water in the tubes was exposed to hot and cold zones between successive fins, so that ice was formed and melted as the water moved up the tubes, thus producing the same result as described above. Although the process was found technically feasible, it was concluded after a long series of experiments that the cost of freshwater produced by a plant of this type would be eight to ten times greater than the cost of water from a distillation plant of comparable size.

II. HYDRATE SEPARATION

Hydrate separation was thought possibly to be superior to freeze separation because crystalline hydrates are formed at temperatures closer to ambient conditions than that at which ice is formed. As pointed out in Chapter 4, hydrate formers, such as methyl bromide, propane, and refrigerants such as Freon-12, form crystals which contain a high percentage of water at temperatures considerably higher than the freezing temperature of water, and are therefore less subject to the in-leakage of heat from the surroundings than is ice. It was reasoned that the energy required would be reduced. Two pilot plants were designed and tested using different hydrate formers — one using Freon-12, CCl_2F_2 [12], and the other, propane, C_3H_8 [13]. While both of these plants produced freshwater, neither one was found to be sufficiently promising from an economic viewpoint to complete the development work. Any possible saving in energy was overshadowed by the increased complication of the process.

The concept of the hydrate-separation process is quite similar to the refrigerant flash vapor compression process, already described in the freeze-separation section. Precooled saline water is pumped into a reactor vessel, where pressure and temperature are controlled so that the incoming hydrate former will combine with the water to form the desired hydrate. Only a small percentage of the hydrate former is used for this step, the remainder being vaporized as the refrigerant for the cooling process, and later compressed prior to its part in the melting process. The hydrate-brine slurry that is formed in the reactor vessel, being a watery mass of very fine crystals, is moved to a filtration station where most of the brine is filtered away before passing the drained slurry on to the counterwasher. The washed hydrate is then pumped into a decomposer vessel where it is brought into contact with the compressed vapor removed from the reactor vessel. Here the hydrate is melted, and the combined melt water and liquid hydrate former then pass on to a decanter where they are separated by gravitational methods. Some 10% of the product water is returned to the counterwasher and the remainder is removed as useful product. The liquid hydrate former is returned to the reactor vessel and the brine is removed from the plant at the exit of the filtration portion of the process. It will be noted that by using the compressed excess refrigerant, removed from the decomposer in the dissipation of heat leakage and energy input along with a final heat sink of cooling water, it was possible to eliminate the separate refrigerator system of the VFVC freeze-separation system that is shown in Figs. 11-3 and 11-4. It will also be noted that, due to the very small size of the hydrate crystals formed under the conditions used in the hydrate-separation pilot plants, this process involves a filtration which is not present in the freeze-separation process. Many technical problems remain to be solved before the hydrate separation process can be regarded as fully developed, even in the pilot plant stage.

REFERENCES

1. A. Peled, Operation of the Freeze Desalination Plant at Eilath, Israel, Proceedings First Intern. Symp. on Water Desalination, Oct. 1965, Vol. 3, pp. 1-10.
2. R. Consie et al., Vacuum Freezing Vapor Compression Process: One and Five Million Gallon per Day Desalting Plants, Res. Develop. Progr. Rept. No. 451, Office of Saline Water, U.S. Dept. Interior, May 1969, 67 pp.
3. M. Pachter and A. Barak, The Vacuum Freezing Vapor Compression (Zarchin) Process: Present Status and Future Trends, Desalination, 2 (No. 3), 1967, pp. 358-367.
4. H. M. Hendrickson and R. W. Moulton, Research and Development of Processes for Desalting Water by Freezing, Res. Develop. Progr. Rept. No. 10, Office of Saline Water, U.S. Dept. Interior, 1956.
5. R. Consie et al, Vacuum Freezing Vapor Compression Desalting Process, Res. Develop. Prog. Rept. No. 295, Office of Saline Water, U.S. Dept. Interior, 1967, 99 pp.
6. C. M. Bosworth et al., A 15,000 gpd Freeze-Separation Pilot Plant for Conversion of Saline Waters, Advan. Chem. Ser. No. 27, 1960, pp. 90-104.
7. A. Zarchin, Demineralizing of Sea Water by Freezing, Israeli Patent No. 7764, 1953.
8. Hittman Associated, Inc., Vacuum Freezing Vapor Compression Desalting: State of the Art (1968), Res. Develop. Progr. Rept. No. 491, Office of Saline Water, U.S. Dept. Interior, Oct., 1969, 57 pp.
9. A. Barduhn, The State of the Crystalization Processes for Desalting Saline Waters, Desalination, 5 (No. 2), 1968, pp. 173-184.
10. N. Ganiaris et al., Secondary Freezing Desalting Process: Operation of a 15,000 gpd Pilot Plant, Res. Develop. Progr. Rept. No. 416, Office of Saline Water, U.S. Dept. Interior, March, 1969.

11. R. C. Himes et al., Evaluation of the Zone-Purification Process as a Method of Preparation of Potable Water from Sea Water, Res. Develop. Progr. Rept. No. 36, Office of Saline Water, U. S. Dept. Interior, May, 1960.
12. K. Garrison et al., Freon-12 Hydrate Desalting Process. Res. Develop. Progr. Rept. No. 368, Office of Saline Water, U. S. Dept. Interior, August, 1968.
13. V. C. Williams et al., Development of Propane Hydrate Desalting Process, Res. Develop. Progr. Rept. No. 373, Office of Saline Water, U. S. Dept. Interior, August, 1968.

Chapter 12

ION EXCHANGE

All of the desalting processes dealt with so far have utilized phase changes to remove freshwater from saline solutions. The basic requirement in these processes is that the latent heat for the phase change must be transferred into and out of the product water. Consequently, the energy requirements are essentially independent of the initial salinity of the water source. In contrast to this situation, there are several processes in which the energy requirements are in proportion to the amount of salts which must be removed. In these processes, which include ion exchange, electrodialysis, and reverse osmosis, the unwanted ions are removed from the product water under conditions which are approximately isothermal at ambient temperature, by using energy in forms other than thermal, i.e., mechanical energy, electrical energy, and chemical energy. When the amount of salts to be removed is only a small fraction of the water to be recovered, it is conceivable that the amount of energy required by these processes will be less than that required for distillation. This has proved to be true for initial salinities less than about 5000 ppm, thus increasing primary interest in this field. The relative newness of these processes, together with the great effort being expended toward their development, may lead to future applications to the desalination of seawater, where presently distillation is more economical than these newer methods.

The first of these processes to be considered is that of ion exchange, so selected because a basic understanding of the materials involved will contribute to the explanation of all the other processes.

As indicated in Chapter 4, ion exchangers are solid materials which contain loosely bonded, inoffensive ions. When these materials are contacted with saline waters containing offensive dissolved ions, there is an exchange of ions, by which the unwanted dissolved ions are absorbed by the exchanger material and the inoffensive ions go into solution. When a softening process is involved, the inoffensive ions in the exchanger material are generally highly soluble ions, such as sodium, so that, although the process changes the nature of the chemical content, it changes the total dissolved content very little. On the other hand, when desalting is the object, the ions on the exchanger material must be either hydrogen and hydroxide ions, i.e., water-forming ions, or ions which can be separated from the solution by precipitation or gasification.

I. APPLICATION OF ION EXCHANGE

The use of ion exchange for desalting has long been applied to the deionization of water needed for precise chemical manipulations and for boiler feed in nuclear power plants. Because the amount of water required and the quantity of salts to be removed were relatively small and the need for extreme purity so very great, ion exchange was considered warranted in these cases, whatever the cost. However, when it became important to process greater quantities of water by this method, new efforts were made to reduce the excessive costs, particularly by employing recoverable regenerants to renew the effective chemicals of the exchangers. At first it was deemed essential that the exchangeable ions left on the ion-exchange material after regeneration be hydrogen and hydroxide ions; hence, the regenerants in the early days were limited to acids and bases. The high costs of these regenerants have, in the past, been the chief limiting factor ruling out the direct ion-exchange process for large-scale water production. As a consequence there has been considerable research directed toward finding new regenerants that can be both effective and less costly.

An early example of unregenerated ion exchange is that of the emergency water desalter developed during World War II for use on life rafts by personnel from wrecked ships and aircraft, in which seawater was purified by precipitation of the offensive salts. This device consisted of a watertight bag containing crushed pellets of silver zeolite and barium hydroxide. When the bag was filled with seawater and was allowed to stand long enough for an exchange to occur between the silver ions on the zeolite and the sodium ions of the seawater, the sodium ions became attached to the zeolite, precipitating the relatively insoluble silver chloride. At the same time the barium ions from the barium hydroxide of the exchanger combined with the sulfate ions in the seawater to precipitate barium sulfate. As a result, the ions remaining in the seawater to any significant extent were chiefly calcium and magnesium and the salinity was reduced to about 5000 ppm. This reduction, although not ideal, was enough for the human blood stream to tolerate, and did not cause dehydration as would have resulted from the ingestion of untreated seawater. A filter screened the precipitate, thus allowing the processed water to be sucked from the bag.

The serious limitation of this exchanger device arose from the fact that silver zeolite could not be regenerated by the only available regenerating material, i.e., seawater. Because of this, additional supplies of silver zeolite were needed for each processing of drinking water. In spite of the high cost of the silver zeolite, it was justified because of the emergency nature of its usage. It can be seen, however, that without the recovery of the silver ions this scheme would be too expensive for large-scale desalination.

Over the years several schemes have been proposed for adapting ion exchange for large-scale production. These include methods which involve (1) regeneration by using ions which can be recovered from the product water as gases, (2) regeneration of ion exchangers by heating, and (3) selective withdrawal of ions. For an understanding of these schemes it is essential to consider the nature and properties of ion-exchange materials.

II. ION-EXCHANGE MATERIALS

A. Natural Zeolites

The oldest application of ion exchange is that of water softening, i.e., the replacement of calcium and magnesium ions in natural waters with sodium ions from an ion-exchange material. While materials other than zeolites are now used more generally for water softening, a description of zeolites and their properties has basic importance.

Zeolites occur in nature as crystalline solids in the form of rocky deposits, generally associated with basic lavas, such as basalt, in areas where volcanism has been active. There are numerous chemical forms of zeolite, but all are alumosilicates. Structurally, the zeolite crystal consists of SiO_4 and AlO_4 groups, each group arranged tetrahedrally with the oxygen atoms shared between pairs of groups. Since the valence of silicon is four and that of aluminum is three, the aluminum group has a net negative unit charge. This negative charge attracts and holds positive ions, such as sodium, within the crystal lattice. These positive cations are not fixed in position, but move within the lattice structure and can be replaced by other cations capable of entering the structure. Because of the negative characteristic of the zeolite, it is useful for cation exchange only.

The crystal structure of a zeolite is conceived as being quite regular with its outer surface containing openings or pores between the tetrahedral groups through which ions, if they are small enough, can enter or leave the crystal. The dimensions of these openings are small, about 3 to 9 Å in average width, so that larger ions and molecules are excluded. A further characteristic of zeolites is that they are soluble in acid, and therefore cannot be regenerated by acid to utilize hydrogen ions. Because of this fact there have been many attempts to synthesize crystalline exchangers from inorganic materials, based upon the above observed structure of zeolites, but with characteristics capable of being regenerated. Numerous studies have been carried on in many fields other than water desalting, details of which can be found in books such as those by Helfferich [1] and Kunin [2].

B. Synthetic Resins

Recent development of plastics has made possible the production of polymerized organic networks having properties similar to those of zeolites. Generally referred to as ion-exchange resins, these materials are gels rather than crystals. They consist of an elastic matrix made up of hydrocarbon chains which have been linked together by polymerization and which contain inorganic ions with excess charges. These excess charges are analogous to the charges due to the valence differences in zeolites, and in a similar way attract and hold counter ions within the gel structure. Also, the charged sites within the gel can be positive as well as negative, and depend for their signs on the particular inorganic ion attached to the hydrocarbon units. For cation exchangers the negative charges on the matrix can be furnished by ions such as SO_3^- and COO^-, the former resulting in an exchanger whose capacity in insensitive to pH. The latter one combines with hydrogen at low pH to form undisassociated COOH and thus is active only at high values of pH. Similarly, anion exchangers may contain pH-sensitive NH_3^+ as the attached ion, or $N^+(CH_3)_3$ which remains ionized even at high values of pH.

1. Production of Synthetic Resin Beads

There are a number of polymers suitable for the matrices of ion exchangers, several of which are described by Helfferich [1]. One of the most common is polystyrene, which can be used both as a cation and an anion exchanger. Styrene monomer (vinyl benzene) is mixed with a small amount of divinyl benzene and the mixture polymerized in the form of small spherical particles or beads. Variously designated as bead, pearl, or suspension polymerization, the process for producing these beads is similar to that used for preparing plastic stock for molding. The monomers, in this case a styrene and divinyl benzene (generally a combination of divinyl benzene isomers and ethyl styrene), are mixed together with a catalyst such as benzoyl peroxide. This mixture is then added to a thoroughly agitated aqueous bath which is maintained at a temperature between 85° and 100°C, the temperature range appropriate for polymerizing the styrene. Agitation of the bath causes the mixture to break up into small droplets, and surface tension draws each droplet into a spherical shape. A stabilizer, such as gelatin or magnesium silicate, can be added to the bath to aid in controlling droplet size, which is affected primarily by the viscosity of the bath and its agitation. By controlling these three factors a fairly uniform range of bead sizes can be obtained. Different combinations of viscosity, agitation, and stabilizer can be used to obtain bead diameters from 0.1 to about 5 mm, the range which has been found to be preferable for ion exchange.

The beads produced, as described above, are insoluble in water, but will swell in water by an amount determined by the degree of crosslinking of the polystyrene. Since polystyrene itself is a thermoplastic chain polymer, it has no crosslinking. However, the inclusion of the divinyl benzene as a copolymer provides the crosslinking feature and results in a highly stable thermosetting material in the form of an elastic gel. The amount of crosslinking depends upon the amount of divinyl benzene in the mixture, this amount varying from 5 to 12 mole % in ion-exchange resins. The larger percentages of divinyl benzene produce more rigid structures. A very common percentage of divinyl benzene in ion-exchange resins is 8.5%.

The above production scheme results in resin beads which are essentially neutral, so far as electrical charges are concerned. It is necessary, therefore, to activate these sites by chemical reaction. For cation exchangers this is accomplished by sulfonation, i.e., by immersing the resin beads in concentrated sulfuric acid, which causes one sulfonic acid group to attach itself to each benzene ring in the polymer. This gives a multitude of negative fixed charges to the polymer matrix, so that an equivalent number of cations, e.g., hydrogen ions, can be held within the boundaries of the matrix. Then, when immersed in an aqueous solution containing other cations in appreciable concentration, the solution cations will enter the matrix and displace the hydrogen ion.

For anion exchangers, the polystyrene beads are first subjected to chloromethylation and then treated with ammonia or with alkyl amines. When trimethylamine is used, for instance, the resultant exchanger can be thought of as having the following group attached to many of the benzene rings:

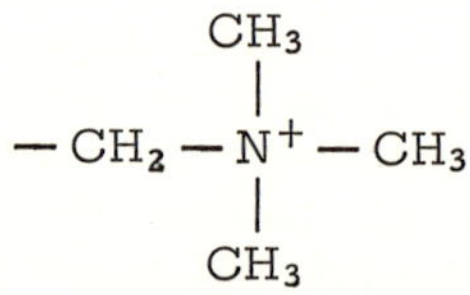

This combination will give a net positive charge to the resin and enable it to hold an equivalent charge of anions within the matrix. The number of active sites in anionic resins does not generally equal the number of benzene rings.

The resins noted above are strong-acid and strong-base resins. They have been produced by many manufacturers and are used widely. These manufacturers, however, use many other materials and schemes for producing resin beads and granules, and make available special purpose resins as well as resins spanning the entire range from strong-acid to strong-base types. A partial list of some of the more common ion-exchanger materials is given in Table 12-1.

2. Properties of Ion-Exchange Resins

To be useful for water purification, ion-exchange resins must possess certain chemical and physical properties. From a chemical viewpoint, a resin should have a maximum exchange capacity and should be stable when exposed to waters containing commonly encountered ions and gases. Also it is important to know its possible selectivity characteristics for various ions. Physically, an ion-exchange resin is characterized by the particle size (average diameter,

TABLE 12-1

Ion-Exchange Materials

Matrix	Form	Strength	Functional groups[a]
Anionic resins			
Polystyrene-DVB	Beads	Strong	Quaternary ammonium*
			Sulfonium
			Phosphonium
Phenol-formaldehyde	Granules	Weak	Primary amine, Secondary amine, Tertiary amine } *
Epoxy-polyamine	Beads or granules	Intermediate	Tertiary amine, Quaternary ammonium } *
Cationic resins			
Polystyrene-DVB	Beads	Strong	Nuclear sulfonic acid
			Methylene sulfonic acid
Polystyrene-DVB	Beads	Intermediate	Phosphonous acid
			Phosphonic acid
Phenol-formaldehyde	Granules	Intermediate	Phosphoric acid
Acrylic	Beads	Weak	Carboxylic acid*
			Phenolic hydroxyl

[a] Resins most commonly used are indicated by an asterisk.

maximum and minimum), the dry density, the moisture content, and the rinse requirement. While values for most of these properties can be obtained from the manufacturer for any particular resin, some general comments may be helpful.

a. Exchange Capacity. The exchange capacity of an ion-exchanger material is expressed in terms of the number of equivalents of exchangeable ions to be found in a unit volume of exchanger. Two values of capacity are important, namely, the total capacity and the "breakthrough" capacity. The total capacity is the number of equivalents which could be exchanged if every free ion were utilized. The total capacity is a true property of the resin and is independent of the type of bed or flow regime imposed on the resin. In contrast, the "breakthrough" capacity is the number of equivalents of ions exchanged with the saline water up to the moment when the concentration of the unwanted ions in the product effluent reaches some predetermined small increment above zero salinity. This capacity measure will depend on the bed arrangement and the flow conditions, and is hence not a constant for a given resin. For comparison purposes this capacity must be measured under certain specific conditions.

To determine the total capacity of a given resin, the exchangeable ions within the resin must be withdrawn under conditions favorable to the exchange, and their quantity measured, the procedures followed varying with the strength of the ions. For the case of a strong-acid cation exchanger, its exchange capacity can be measured in terms of its salt-splitting capability. To accomplish this, a given volume of resin, e.g., 60 to 80 ml, is preconditioned by cycling two or more times, first with concentrated hydrochloric acid and then with concentrated sodium chloride, repeating to finally leave the resin in the hydrogen ion state. The excess acid from this final cycling is rinsed from the bed with distilled water until the pH of the rinse water at outlet is neutral. The bed is then exhausted with an excess of the concentrated sodium chloride solution, followed by a distilled water rinse. The combined effluents of these two last flows is then titrated with sodium hydroxide to determine the amount of hydrogen ion which was removed from the bed. The ratio of the total gram equivalents of hydrogen ion removed to the volume of the exchanger material involved is the measure of total capacity.

This exchange capacity can be estimated when certain assumptions are made with respect to the number of exchange sites per molecule of the matrix. For example, a cation exchanger made by sulfonation of a polystyrene matrix may be expected to have one exchange site for each benzene ring, i.e., for each styrene molecule. With the formula $C_8H_8O_3S$, for instance, the molecular weight of the sulfonated molecule is 184.2. Thus the exchange capacity is 1.00/184.2 or 0.00543 equiv/g, usually expressed as 5.43 meq/g. However, an ion-exchange material is seldom a simple polystyrene matrix, but to it is added a crosslinking copolymer such as divinyl benzene and ethylstyrene which introduces units with higher molecular weights, namely, 210.2 and 212.2, respectively. These higher molecular weights have the effect of reducing the exchange capacity per unit weight. For instance, a resin with 8% divinyl benzene would have a capacity of 5.35 meq/g of dry weight instead of the above

5.43. To convert this figure to a volumetric basis, the weight per unit volume must be known. For example, a specific resin may have a weight per unit volume value of 501 lb/ft^2, or 800 g/liter. If this resin's bulk density is measured when it is saturated with water, i.e., 53% of water per weight of the above particular resin, the total capacity of this resin would be 5.35 times 800(1 - 0.53) = 2012 or 2.01 equiv/liter. This order of magnitude is typical of sulfonated polystyrene cationic resins. The exact magnitudes vary with the bead size and the amount of crosslinking. It has been found that the extreme variations of total capacity of sulfonated polystyrene crosslinked with divinyl benzene made by several manufacturers range between 1.2 and 2.5 equiv/liter. The total capacities of cationic resins produced from other materials vary widely, with values as low as 0.33 and as high as 4.2. Values of total capacity for styrene-based anionic resins are generally somewhat less, the highest value of capacity quoted by resin manufacturers being 1.6 equivalents per liter. For all anionic resins the total range of capacities is 0.7 to 3.0 equiv/liter.

b. Selectivity. The most pronounced type of selectivity encountered in water treatment is that of electroselectivity, which is characteristic of many resins and means that ions in a mixture will be absorbed by an ion exchanger in the order of their valences, with ions of the higher valences being absorbed first, i.e., calcium will be absorbed preferentially to sodium, and sulfates to chlorides. This is not universally true, however, since the ions absorbed must also be small enough relative to the pores in the exchanger surface to enter the exchanger matrix. The alkali cations have additional selectivity characteristics in that they are absorbed roughly in the order in which they appear in the Periodic Table. This order corresponds to the diameters of the hydrated ions. Since the purpose of desalination is to reduce all salts, the order of their reduction is relatively unimportant in ion-exchange water treatment, so no further details will be presented here, but may be found in the reference cited [1].

c. Physical Properties. Ion-exchange resin beads are generally sized using sieve techniques to give a certain range of particle diameters. For example, a sieve specification of 10-50 (U.S. Standard) is a mixture of bead diameters ranging from 0.297 to 2.00 mm; a specification of 18-50 is a mixture of bead diameters of 0.297 to 1.00 mm. In specifying sieve sizes it is essential to know the sieve standard used, since there are several systems, the sieve sizes of which are not necessarily comparable. The U.S. Standard sieve series are measures of both fractional numbers of meshes per lineal inch and also integral numbers of millimeters and the common fractions of 1 mm, such as $\frac{1}{2}$, $\frac{1}{4}$, and $\frac{1}{8}$. This is slightly different from the British Standard in which sieve sizes are based on integral numbers of meshes per inch. In addition, there are some numbers which correspond to integral numbers of millimeters and their fractions. The Tyler Standard is based on integral numbers of meshes per inch, but has no numbers which correspond to integral numbers or fractional numbers of millimeters.

III. ION-EXCHANGE COLUMNS OR BEDS AND THEIR OPERATION

The application of the ion-exchange principle to water desalting is usually accomplished in ion-exchange columns or beds, similar to arrangements in sand filters. In these beds the ion-exchange particles occupy a space up to 4 ft in height on top of fine mesh screens or beds of sand which prevent the loss of the small ion-exchange particles. Piping is arranged to distribute water over the top surface of the ion-exchange bed during the service, regeneration and rinse portions of the operating cycle, and to remove the resulting flows from the bottom of the bed. Piping is also provided to distribute water over the bottom surface of the bed during the backwash portion of the cycle, and to remove the backwash effluent from the top of the bed.

With a bed arranged as above the complete operating cycle includes four periods:

1. Service period, with downflow of water during which saline water is desalted. Flow rates average 2 gpm/ft^3 of resin.

2. Backwash, during which water is pumped upward through the resin to remove any foreign matter which may have entered and to reclassify the resin particles. Flow rates for this purpose vary between 4 and 6 gpm/ft^2 of bed cross section.

3. Regeneration period, during which the regenerant solution is passed downward through the bed, displacing the unwanted ions from the exchanger and restoring it to its original condition. Flow rates for this process are less than for the service period and vary from 0.5 to 1.5 gpm/ft^3 of resin.

4. Rinse period, during which the spent regenerant remaining in the pores of the bed is displaced. The water used for this purpose can be the saline feed, and is discarded as long as the effluent shows contamination with the spent regenerant. The service period begins again when a monitor in the effluent stream, measuring either pH or electrical conductivity, indicates an acceptably low salinity.

The above described cycle results in the service period occupying only part of the total operating time, so that a single bed would not give a continuous flow of product water. Thus, in production plants where continuous flow is desired there must be several ion-exchange beds, each of which is capable of giving the required flow rate. This multiplicity of beds represents a large investment and is a major cost component of the water being desalted. Because of this inefficiency of "fixed beds," there have been several suggestions to counteract this deficiency by designing "moving beds" in which the stages of the above cycle would affect only small continuing portions of the bed. In this manner a large fraction of the resin would be active most of the time. Weiss [3], measuring the resin productivity in pounds of product per day per cubic foot of resin, achieved substantial increases in productivity by using such "moving bed" schemes. Obviously this would apply only to large plants, for the cost of the equipment and its controls would be too great for practical application in small plants. Continuous ion-exchange equipment has been developed for water softening, for obtaining magnesium from seawater, and for various large-scale metallurgical operations. It is argued, therefore, that because a large fraction of the cost of producing freshwater by ion exchange is due to resin costs, any increase in productivity arising from a more continuous operation of the equipment could measurably reduce water costs in large-scale plants.

With the operation of the several aforementioned cycles in a practical plant, it has been found that only a portion of the resin total exchange capacity is utilized during the service period. The nature and reason for this low capacity can best be understood by considering a fully regenerated ion-exchange column or bed through which is passed a volume of saline water somewhat greater than the ion exchanger's total capacity to deionize. The volume of saline water containing ions equivalent to the total exchange capacity of the exchange materials in the bed is plotted by the dotted line in Fig. 12-1 and is indicated by point A. The vertical line above point A indicates the instantaneous rise in salinity to the influent value, if the bed could be operated to completely deionize this theoretical amount of water. In actual operation, however, when the salinity of the effluent product water is plotted against the accumulated volume of water which has been passed through the bed, the curve follows the solid line of Fig. 12-1. It is noted that the first part of this curve is horizontal at zero effluent salinity, indicating for that quantity of volume complete deionization of the saline water. At a certain volume, 30 to 80% less than the total capacity, and indicated by point B, the curve breaks from zero salinity and begins to rise, continuing upward until it approaches asymptotically the influent salinity of the water. Point B is called the "breakthrough" point and

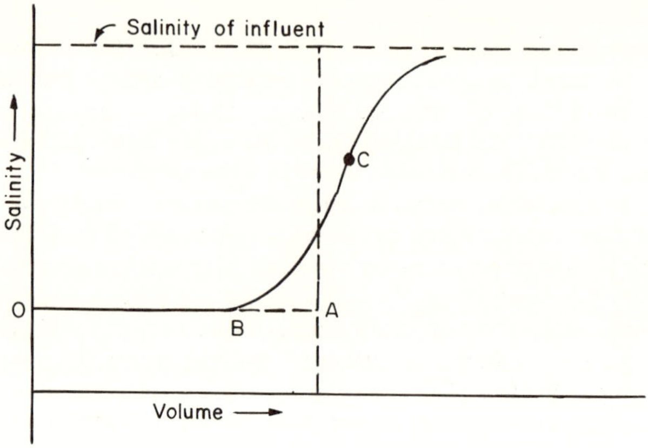

FIG. 12-1. Breakthrough curve for fixed-bed ion exchange (schematic only).

marks the increase in salinity resulting from the "breaking through" of one or more of the undesired ions. For the production of highly pure water this point limits the volume of product which can be collected from the ion-exchange bed before regeneration. For the production of domestic and general industrial water, the acceptable concentration of which has been set by the United States Public Health Service standards at a level as high as 500 ppm, the volume of the product effluent that can be collected before regeneration will be at some point beyond the "breakthrough" point, e.g., C. This point represents the volume of water whose weighted average salinity is equal to an acceptable salinity.

The location of the "breakthrough" point for ion-exchange treatment of a given saline water is influenced by many factors, including the design of the bed as well as various operating conditions. Design elements, such as bed height, ratio of wall perimeter to bed surface area, and uniformity of water distribution over the top surface of the resin bed, must be optimized to achieve best results. The operational features which influence the location of the "breakthrough" point are the flow rate through the bed, the temperature of the water, and the sizes of ions to be removed. Increasing the flow rate through the bed moves point B to the left because of the decrease in time allowed for ionic diffusion into and out of the exchanger particles as well as the increase in the tendency for water to "channel" along the wall surfaces. Increasing the temperature of the water increases the rate of ion diffusion and thus moves point B to the right.

IV. ION-EXCHANGE PLANTS FOR DESALTING WATER

A. Plants Producing High-Purity Water

The water used for producing steam in modern power plants, especially those using nuclear reactors as heat sources, and in some manufacturing establishments and laboratories, must be quite free of dissolved materials to prevent corrosion and salt deposition. In such plants an ion-exchange unit is placed in the feedwater system to remove any ionized materials which may have entered the feedwater in spite of the many precautions taken to avoid such contamination. It is necessary, however, to recognize the basic distinction between this type of ion exchange for high-purity polishing of water relatively low in salinity, and the type of ion exchange for producing acceptably pure water from saline sources. The small quantity of water required in the operation of the former and the low percentage of salts in its solution involves relatively small amounts of ion-exchange chemicals, and the high priority requirements for purity justify any necessary expensive procedures. Because the quantities involved are so small, capital cost and not regenerant cost is the significant factor. In the case of desalting saline water for potable use, on the other hand, the overriding consideration is the recurring regenerant cost. For the above reasons, therefore, it is feasible in plants producing high purity water to use a straightforward ion-exchange method in which acids and bases, even though exceedingly costly, provide the hydrogen and hydroxide ion regenerants in the exchange beds. Through their use the exchange process results in the formation of water molecules to take the place of the unwanted ions, and to thereby furnish the high purity water desired.

When this scheme is considered for desalting a municipal supply, the cost of the hydrogen and hydroxide ions needed for regeneration becomes a major component of the probable expenditure for operating an ion-exchange plant, and their costs render such regeneration impractical for the desalting of saline water. For example, using the above process to desalt water having the composition of the Coalinga water, as given in Chapter 2, would entail the expenditure of over 20¢ per kgal of product water for the sulfuric acid to replace the hydrogen ion and about twice as much for replacement of the hydroxide ion.

B. Plants for Desalting Saline Water

Because ion-exchange processes have many desirable features for water desalting installations, there have been a number of attempts to substitute less costly regenerants than the above hydrogen and hydroxide ions in order to make the process acceptable for desalting application. Although there are many large ion-exchange plants in municipal water supply systems, all of these are used for the purpose of softening low salinity water rather than for desalting saline water. Again, such a softening operation must be distinguished from a desalting process, in that the former process does not remove the ions from a solution, but merely replaces the undesirable calcium ions with sodium. A desalting device must find means of separating and discarding the unwanted ions from the product water. The success by which cheap and plentiful sodium chloride has been used in the softening process has stimulated a search for cheap regenerants for the desalting operation. In an effort to remove the regenerant ions from the product water, this search has resulted in several proposals using various chemicals which can be precipitated or gasified chiefly by the application of heat, and thereby complete the desalination process.

1. Use of Heat for Regenerant Recovery

One attractive scheme for regenerating both the anionic and cationic resins by the use of ammonium bicarbonate has been suggested by Gilliland [4]. His general proposal was to regenerate a mixed-bed exchanger with ammonium bicarbonate. The product water passing through this bed would emerge laden with ammonium and bicarbonate ions in place of the original unwanted ions. Before being discharged from the plant this product water would be heated to drive off the ammonium bicarbonate, in the form of NH_3 and CO_2 gases. These gases would then be collected and redissolved to furnish the major part of the regenerant, "makeup" ammonium bicarbonate supplying the balance. This proposal was thoroughly investigated by Stewart [5], who found that the exchange of ions could be successfully accomplished by this method but that the recovery of the gases required as much heat energy as though the original water had been distilled. He points out, however, that since the amount of material to be vaporized was only a small fraction of the weight of the product water, it could have economic possibilities if the heat energy were applied as efficiently as in distillation.

A related proposal by Ranck [6] involves the use of calcium bicarbonate as the regenerant in a series of ion-exchange beds. In his scheme the unwanted ions of seawater are exchanged with calcium and bicarbonate ions from the exchanger beds, producing a solution of product effluent with only calcium bicarbonate in ionized form, which, when heated in a vacuum precipitator, causes the formation and precipitation of calcium carbonate. This latter material is filtered, leaving pure water. Calculations made by Ranck indicate that five stages of this treatment can reduce the sodium chloride content of seawater to about 170 ppm. Although he was able to accomplish the desired exchange in the above operations, the estimated quantity of heat required to precipitate the calcium carbonate in the several stages was so large that it made its economic feasibility questionable.

A third possible scheme for using heat in the recovery of regenerants was studied by Busch and Lynn [7], who proposed the use of a strong-acid resin and a weak-base resin for the exchanger beds. Regenerants for these would be hydrochloric acid for the cation resin and ammonium hydroxide for the anion resin, both of these materials being low in price. The use of these regenerants makes hydrogen and hydroxide ions available for exchange with the unwanted ions in the saline water. In order to partially recover the acid and base from the regenerant waste, which is largely ammonium chloride, magnesium oxide is added to the solution, thereby reacting with the ammonium chloride to become magnesium chloride and in the process releasing ammonia gas, which is collected and dissolved for reuse as ammonium hydroxide. The remaining waste, magnesium chloride, is then subjected to heat and evaporated to dryness, becoming in the process magnesium oxide and releasing hydrogen chloride gas. Through these complicated steps there is accomplished a partial recovery of the acid and base, and of the active agent, magnesium oxide. Laboratory tests have indicated that this process should use less heat than the others.

2. The Use of the SUL-biSUL System

This scheme for desalting brackish water, as patented by Elgin International [8], uses two ion-exchange beds in series. The first bed is regenerated with hydrogen ions and removes cations from the saline water. The second bed is regenerated with sulfate ions, and the effluent from the first bed, charged with hydrogen ions, changes the sulfate ion to bisulfate ion without removing these ions from the exchanger material, and thereby releases one-half of the exchange sites so that they can be occupied by the anions contained in the saline water. With the passage of the saline water through these two beds most of the unwanted ions will have been removed. The hydrogen ion regeneration of the cationic bed is accomplished in the usual manner with sulfuric acid. The anionic bed, in which one-half of the sites are occupied by bisulfate ions, can be regenerated to sulfate ions by charging the bed with either sulfuric acid or the saline water itself which is usually high in sulfate concentration. The chief claim of this process is that the expensive hydroxide ion regeneration is replaced by the much less expensive sulfate regeneration.

3. Use of Electricity for Regeneration

The use of dc electricity for the direct regeneration of ion exchangers has been tried, based on the assumption that the salinity of most natural waters will furnish the conductivity to set up the electrical current and thus permit the ions in the exchanger material to go into solution. Any application of this scheme, however, has been very inefficient and is not in use.

A somewhat different concept of the use of electricity for ion-exchange regeneration has been proposed by Murphy [9], who has been working for some years on the use of chemically modified carbon electrodes to selectively absorb ions which are delivered to them by the flow of electric current. These electrodes become "saturated" when the ionic sites are used up, after which they are reactivated by reversing the current to drive the ions back into the stream. The service period of product flow is that during which the ions are being absorbed on the carbon electrodes and the regeneration period is the reverse electric current flow. In the process of making the electrodes, a thick sludgelike slurry of carbon is prepared and worked into an inert fibrous backing material, after which the assembly is ovendried and calendared to improve conductivity. Materials used to improve the ionic absorption are similar to those used as functional groups in ion-exchange resins, and are added to the slurry before the electrode is formed. Although this process has been under development for several years, it is still regarded as being in the experimental pilot plant stage.

V. SOFTENING OF SEAWATER

The use of ion exchange in the pretreatment of seawater in order to prevent scale formation in distillers has already been described in Chapter 5, so very brief mention of the basic concept will be considered here. The softening process is directed to the reduction of the calcium ion content, thereby lessening the possibility for calcium sulfate scale to form. The regenerant employed to accomplish this reduction, assuming that its sodium chloride salinity is great enough, is the waste brine from the distiller. Studies made by Klein [10] indicate that a concentration of seawater brine to about four times that of normal seawater will provide a sufficiently concentrated sodium chloride regenerant to restore the ion-exchange bed to its initial condition.

A similar scheme has been reported by Kunz [11] in which the discharging concentrated brine is used for regeneration in a continuous ion-exchange process. In this version, there are two columns, in one of which, the product column, the incoming seawater passes countercurrently to the moving ion-exchange material and exchanges calcium for sodium ions from the exchanger beads. This water, now low in calcium ions, is passed on as distiller feed. The exchange material leaving this column as a fluidized stream enters the second, or regenerant column, where it moves countercurrently to a stream of the concentrated brine which is being discharged from the distiller. Sodium ions from this brine exchange with the calcium ions on the exchanger beads, so that the exchanger material leaving this column is regenerated with sodium ions and is ready to be returned to the product column, still in the fluidized form. The brine discharging to waste from this column carries with it the unwanted calcium ions. When this scheme is used as a softening process for distiller feed it has been found that brine temperatures have reached as high as 320°F in the distiller and that no scale formation has occurred under these conditions.

REFERENCES

1. F. Helfferich, Ion Exchange, McGraw-Hill, New York, 1962.
2. R. Kunin and R. Myers, Ion Exchange Resins, John Wiley, New York, 1950.
3. D. E. Weiss, The Role of Ion Exchange Desalination in Municipal Water Supplies, Desalination, 1 (No. 2), 1966, pp. 107-128.
4. E. R. Gilliland, Fresh Water for the Future, Ind. Eng. Chem., 47, 1955, p. 2410.
5. P. B. Stewart, Sea Water Demineralization by Ammonium Salts in Ion Exchange, Advan. Chem. Ser., No. 27, 1960, pp. 178-191.

6. J. P. Ranck, A Preliminary Technical and Economic Investigation of Sea Water Demineralization by Ion Exchange for Calcium and Bicarbonate Ions, and Their Subsequent Removal by Thermal Decomposition, Desalination, 6 (No. 1), 1969, pp. 75-86.
7. G. A. Busch and S. Lynn, Deionization of Brackish Water Using Thermally Regenerated Reagents, SWCL Report No. 72-2, Univ. of California, Berkeley, California, Nov., 1972, 64 pp.
8. K. Schmidt et al., SUL-biSUL Ion Exchange Process: Field Evaluation on Brackish Waters, Res. Develop. Progr. Rept. No. 446, Office of Saline Water, U. S. Dept. Interior, May, 1969.
9. G. Murphy and J. H. Tucker, The Demineralization Behavior of Carbon and Chemically Modified Carbon Electrodes, Desalination, 1 (No. 3), 1966, pp. 247-259.
10. G. Klein et al., Design and Cost of Ion Exchange Softening for a 50 mgd Sea Water Evaporation Plant, Univ. of California, Berkeley, California, SWCL Rept. No. 68-2, Sept., 1968, 59 pp.
11. J. K. Kunz, Ion Exchange Pre-treatment for Sea Water Desalination Plants, Desalination, 3 (No. 3), 1967, pp. 363-372.

Chapter 13

ELECTRODIALYSIS

I. APPLICATION OF dc ELECTRICITY TO DESALTING

The cost of desalting water by means of electrodialysis has become competitive with distillation in situations where the initial salinity of the water supply is less than about 5000 ppm. This relatively new method of desalination, which uses direct current electricity to pluck unwanted ions from solutions and thereby reduce the concentration of salts in intially saline water, became a reality in the early 1950s after Juda and MacRae [1] developed ion-exchange membranes. Almost simultaneously the TNO group in The Netherlands developed an effective electrodialysis membrane, using parchmentized paper [2]. As explained in Chapter 4, these membranes are of two types, one of them excluding negative ions but permitting positive ions to pass through the membrane structure, and the other excluding positive ions and permitting negative ions to pass through. Both membranes are relatively impervious to the passage of water, and the small quantity that does pass through is due chiefly to the osmotic effect of the difference in concentration between the solutions on the two sides of the membrane. In electrodialysis units for desalting saline water these two membranes are stacked in pairs, thereby setting up an arrangement in which the alternating compartments between the membranes are, in the one, increased in concentration and, in the other, depleted. Commercial installations for this process were begun about 1955, and have been quite successful in treating moderate quantities of water having

TABLE 13-1

Representative Electrodialysis Plants

Location	Type	Number of units	Capacity (1000 gpd)	Year
Arabia/Saudi		2	115	1961
Bahrein		1	100	1963
Bahrein		1	100	1964
Bermuda		1	25	1966
Italy		4	290	1967
Kuwait		1	240	1963
Libya		4	115	1963
Netherlands, Rotterdam		1	185	
Netherlands, Rotterdam		1	260	
Netherlands, Rotterdam		2	480	1968
Tunisia		1	26	1965
Tunisia		3	37	1965
Tunisia		1	38	1967
United States, Buckeye, Arizona		1	650	1962
United States, Coalinga, California		1	28	1959
United States, Port Mansfield, Texas		1	250	1965
United States, Siesta Keys, Florida		11	1200	1969
United States, Webster, South Dakota		1	325	1962

relatively limited initial salinities. Much of the success of this process has been due to the efforts of an American firm which actively has serviced and improved field units in order to keep them abreast of the rapidly advancing technology. Installations of electrodialysis equipment had a world-wide capacity of about 10 mgd in 1971. Some of the more significant installations are listed in Table 13-1.

The general principle of electrodialysis has been set forth in Chapter 4, so that this chapter is concerned primarily with the quantitative application of the principle in terms of equipment and plant design. However, a brief review is essential for a clear understanding of the apparatus.

The basic idea is that the establishment of a current of dc electricity across a stream of saline water will cause ions to migrate across that stream as carriers of electricity in much the same manner as metallic ions are caused to move through an electroplating solution. If a pair of oppositely charged electrodialysis membranes are so placed in an electrically charged stream that their faces are parallel to the direction of water flow, and so oriented that the ions will pass through these membranes out of that portion of the stream between the membranes, the stream between the membranes will become desalted and the streams on the electrode sides of the membranes will become concentrated with the unwanted ions. Obviously the use of a single pair of membranes would not be economical for desalting since the voltage drop through the water and the membranes would be less than the sum of the voltage drops at the two electrodes, and as a result most of the power input would be expended in overcoming the electrode resistances rather than in desalting water. For this reason all practical electrodialysis desalting devices use many pairs of membranes between a single pair of electrodes, thus dividing the saline water flow into many small streams in which most of the current-carrying ions are trapped in half of these streams and the desalted water in the other half of the streams. Only a small fraction of the ions, namely, those coming from behind the membranes adjacent to the electrodes, reaches the electrode compartments where they must be "plated out," either directly or by substitution. While increasing the number of pairs of membranes between electrodes increases the efficiency of electric current utilization, the practical optimum number of pairs for any given design will be dependent upon problems of the mechanical assembly, of sealing against water leakage, and of insulation and control of large dc electrical voltages. Most designs vary between 275 and 500 pairs of membranes per pair of electrodes.

In the development of electrodialysis desalting devices it has been found that two of the major cost components of product water are those for electrical power and for membrane replacement. The electrical power used, i.e., the product of voltage and current, is dependent on the electrical resistance of the path of electricity through the device, and on the fraction of the current involved in the transport of the ions. If voltage is to be minimized in a device with components of given electrical resistance, the resistance of the water streams must be minimized, i.e., these streams should be as narrow as possible, with membranes very close together. This is one reason that in most practical plants the membrane spacing is at most 1 mm.

The effectiveness with which ion transport accounts for the current used in an electrodialysis desalter is customarily discussed in terms of the current efficiency, sometimes referred to as the coulomb efficiency. The amount of electricity carried by the ions through the membranes is equal to the product of the gram equivalents of salt ions involved and Faraday's constant—96, 500 C per g-equiv. The current efficiency is the ratio of the above quantity per unit time to the actual electrical current used, values of this in practical plants being of the order of 90 to 95% or more.

The second major cost item, that for membrane replacement, depends on the useful service life of the membrane, generally from 3 to 5 years, and on its area, since membrane costs are proportional to area. For membranes of a given useful service life and unit cost in a desalting device of a given capacity, the product water cost item is dependent on the membrane area, and hence on the electrical current per unit area, generally referred to as the current density. For minimum product water cost, the current density should be as large as possible in order to use the minimum membrane area. However, there is a practical limit on the magnitude to which current density can be increased in any given electrodialysis desalting device. For each such device there is a critical current density below which the current efficiency is nearly constant, but above which the current efficiency decreases more and more as the current density is increased. Since a primary factor in determining the critical current density is dependent on polarization, further discussion of the critical current density will be found below. At this point it may be noted that practical plants use current densities up to 30 or 40 mA/cm^2.

II. COMPONENTS OF A CONVENTIONAL ELECTRODIALYSIS STACK

The assemblage of membranes and electrodes, together with the appurtenances necessary for confining and directing the streams of water over the membrane surfaces, is generally referred to as an electrodialysis stack. The membranes with their spacers and gaskets form a series of similar assemblies, each of which is called a cell. The electrodes at the ends of the stack have special compartments in which the necessary oxidation and reduction processes occur. A discussion of these several componenets of the electrodialysis stack will aid in understanding its operation.

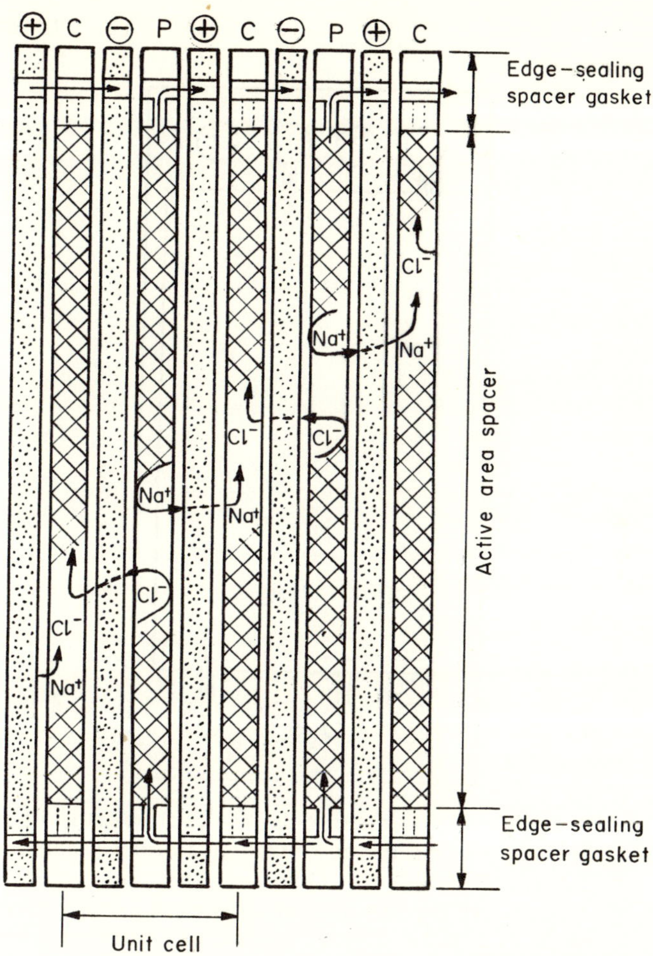

FIG. 13-1. Expanded cross section of electrodialysis cells (schematic only). ⊕ Cationic membranes (negative sites) passing positive ions; ⊖ anionic membranes (positive sites) passing negative ions; P, product compartment; C, concentrate compartment. The repeating "unit cell," designated on the drawing, is the basic cell and consists of a pair of positive and negative membranes together with two spacers and gaskets. The water passages within the cell are indicated, and communicate with channels throughout the entire stack by holes provided, as shown in Figs. 13-2 and 13-3. When assembled the four components of the cell are clamped together to seal against leakage. The product manifolds are shown connected to the product compartments. A second pair of manifolds, located behind those shown, connects to the concentrate compartments, as indicated by the dotted lines.

FIG. 13-2. Sheet flow type spacer gasket (schematic only). 1 and 4, Manifold ports for one stream; 2 and 3, manifold ports for other stream. Water flow between 1 and 4 is shown by arrows. When the spacer gasket is turned over, flow will be from 3 to 2. Spacer thickness equals gasket thickness and occupies entire cutout area of the gasket.

A. Cells

The repeating unit, or cell, consists of a pair of membranes sealed around the edges to prevent leakage, and separated by spacers which hold the membranes apart so as to provide passages for the water to flow across the surface of the membranes. The spacers and edge-sealing gaskets in Fig. 13-1 can be separate or one piece, the latter being preferred by some manufacturers. The membrane pairs are oppositely charged and each one permits ions of only one sign to pass through it. With the ions migrating in the directions shown in the figure, the product and concentrate compartments are as indicated. The flow of current through the stack is horizontal in the plane of the figure and the flow of water in the compartments between membranes is vertical, controlled by the edge-sealing gaskets.

1. Spacers and Gaskets

As indicated in the beginning of this chapter, the basic concept of electrodialysis is that pairs of membranes divide the water stream into many small but parallel streams through which the water flows between the many membranes of the stack. This result is achieved by the design of the edge-sealing gaskets, a representative design being shown schematically in Fig. 13-2. These spacers and gaskets are made of some sheet material, the thickness of this material being equal to the desired membrane spacing. Since the membranes are relatively thin, and therefore flexible, the spacer must contain elements which will prevent adjacent membranes from touching each other and thus short-circuiting the cell. At the same time these supporting provisions must present a minimum of resistance to the flow of water through the cell.

In Fig. 13-2, the gasket is shown as cut out over the central active area to accommodate the porous spacer. There are inlet manifold holes on one side of the gasket and outlet manifold holes on the opposite side, connected to the central cutout portion by narrow slots. It will be noted that the placement of the holes in the gaskets is symmetrical, but that the slots in the gaskets are placed unsymmetrically. The membranes are punched with the symmetrical holes but not the slots, and by proper matching, inlet and outlet manifolds run through the entire stack of membranes and gaskets. Because of the asymmetry of the slots in the gaskets, the alternate product and concentrate compartments are connected by merely reversing alternate gaskets.

There are two general patterns of water flow across the surface of the membranes, namely, sheet flow and tortuous path flow. In sheet flow, the spacer, of the same thickness as the gasket, covers the entire active area, as shown by the cutout portion of Fig. 13-2, and serves the two purposes of maintaining the spacing of the membranes and increasing the turbulence of flow of water next to the membrane surface. Spacers used in sheet flow cells can be sheets of polyvinyl chloride (PVC), perforated and corrugated, or slit and expanded as in metal laths used in the building trades, or open mesh fabrics woven from hard synthetic filaments. While serving the functions noted above, these spacers must also present as little resistance to water flow as possible.

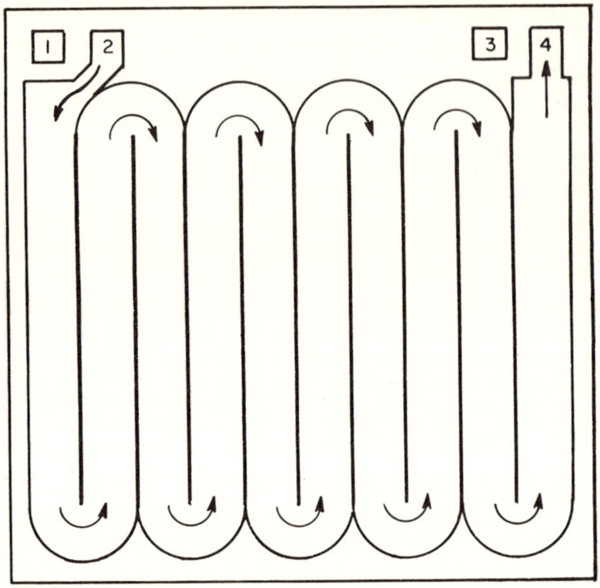

FIG. 13-3. Tortuous path spacer and gasket (schematic only). 1 and 3, Manifold ports for one stream; 2 and 4, manifold ports for other stream. Water flow path shown by arrows for flow from 2 to 4. When the spacer and gasket are turned over, flow is from 1 to 3. Half-thickness tie supports across water flow path are not shown.

The tortuous path type of spacer is usually one in which the spacer and gasket are integral, as shown in Fig. 13-3. In this figure it is seen that the water is forced to flow back and forth across the membrane several times, thus providing a long flow path although the membrane dimensions are relatively small. Not shown in the figure are half-thickness ties across the flow passages which are necessary to prevent contact between adjacent membranes, these ties also serving as turbulence promoters. The example of the tortuous flow path, as shown in the figure, contains five return passes across the sheet and, with a membrane having a return pass length of 3 ft, the five passes would total to a path length of about 15 ft. Because of the long length of its flow path, this type of spacer has the advantage of being able to accomplish the prescribed percentage of desalting with relatively high velocities of water flow within the cell. Greater velocities promote turbulence and reduce the problems of polarization, which, as will be explained in more detail later, is the tendency toward substitution of hydrogen and hydroxide ions from the breakdown of water to replace the dissolved ions of the solution when for any reason there is a shortage of ions to act as carriers of the electricity.

2. Membranes

Since the earlier section assumes the existence of membranes with suitable properties of selectivity and permeability to ions, it is essential to understand the principles involved in their construction in order that the desired properties may be secured. The concept of a membrane with electrically positive or negative sites, capable of being both selective and permeable to ions, is similar to that pertaining to ion-exchange beads. Assuming that gels used for ion-exchange beads could also be formed into sheets or films, it was reasoned that ion-exchange membranes could be produced by polymerization techniques, and charged sites produced by chemical treatment as in the manufacture of ion-exchange beads. However, early attempts to produce membranes by this method were unsuccessful, so the first electrodialysis membranes were made by grinding ion-exchange beads into a powder and dispersing this powder in a plastic binder which could be formed into a sheet. Later techniques made it possible for membranes to be produced in sheets by the above more direct method of polymerization, with modifications to produce the appropriate desired sites.

a. Properties Desirable in Electrodialysis Membranes. Prior to a discussion concerning the manufacture of membranes it is pertinent to review the properties needed in those membranes. Electrodialysis membranes must possess specific mechanical, electrical, and chemical properties, many of which have been measured by standard procedures and ratings. These procedures have been published by the OSW [3] and are in general use for the evaluation of membranes. The nature and significance of these properties are indicated below.

i. Ion selectivity: This is the most important property and is the one which has made electrodialysis possible. The term "permselective" is used to describe these membranes, and indicates that they are permeable to selected ions. Since no membrane is perfectly permeable or selective, the most common measure of permselectivity is the ratio of the potential of an actual membrane separating two aqueous solutions of different but fixed concentrations, to that which would exist for a membrane of perfect selectivity. Permselectivity values for different membranes have been reported to range from 60 to 95%, when tested between solutions having concentrations of 0.5 and 1.0 N KCl at 25°C.

The manner in which membranes are selective in passing ions can be explained in terms of the structure and electrical charge of the membrane. Membranes used in electrodialysis contain charged sites and when not operating are saturated with water containing ions from the solutions on either side. When there is no electrical current flow, the membranes will become electrically neutral by the migration into the membrane of ions of the appropriate electrical sign from the solutions in contact with the membrane. For example, if the charged sites are negative electrically, positive ions will migrate into the membrane, and negative ions will be repelled by the effect of similar charges of the membrane sites. When an electrical potential is imposed on the above membrane, the positive ions already within the membrane together with those from the bounding solutions will act as current carriers, passing from site to site in the membrane structure and producing a continuous migration of positive ions from the solution on one side of the membrane to that on the other. Thus, the presence of negatively charged sites has permitted only positive ions to pass through the membrane. In a similar manner, a membrane with positively charged sites will permit only the negative ions to pass through.

ii. Electrical resistance: The resistance of a membrane to the passage of an electrical current can be compared to that of a solid metallic conductor, and is a function of the geometrical shape and size of the path of electric current flow and of the specific resistance of the material. Using this analogy, although membranes are unlike metallic conductors in that they are porous to water and their specific resistance changes with the water content, the basic resistance of a membrane for any given material will be proportional to the membrane thickness. This would indicate the desirability of using thin membranes.

Because of this difference in resistance due to membrane thickness, it has become customary to use as a basis of comparison the value of "area" resistance. Area resistance is the product of specific resistivity of a given material and the membrane thickness, and is therefore the numerical expression of the resistance of unit area for a given thickness of membrane. In order to minimize the complications of comparison due to ion transfer through the various membranes and the accompanying concentration changes adjacent to the membrane surfaces, this "area" resistance is measured on the basis of ac current at 1000 Hz frequency. Such measurements indicate the resistance of the membrane alone, since the use of ac current prevents ion transfer and the accompanying concentration gradients. Values of area resistance under these conditions have been reported to have magnitudes ranging from 1.5 to 20 Ω-cm^2 in 0.5 N NaCl solutions, depending on the temperature of the solutions, materials of the membrane, etc. It has further been observed that these resistances decrease as salinity increases. While the area resistances determined under the above conditions are useful in comparing the relative desirability of two or more membranes for use with a given type of water, the more truly representative value for effective resistance must be considered equal to the ratio of voltage drop to current density when operating in an electrodialysis cell, because of further phenomena which accompany the ionic conduction of electricity, as will be discussed later.

iii. Mechanical strength: Besides the selective and electrical properties mentioned in the above paragraphs, electrodialysis membranes must have sufficient mechanical strength to withstand the fluid pressures and mechanical forces to which they may be subjected. The fluid pressures involved in an electrodialysis cell are those required to cause flow through very narrow passages. Depending on the cell design, these pressures vary from 15 to 60 psi per stack. To determine measures of mechanical strength, membrane samples may be subjected to the Mullen Burst Strength Test. In this test the membrane sample is placed in a standard testing device in which it is clamped around the edges between two plates, in one of which there is a central circular hole 1.2 in. in diameter. Pressure is applied by pumping glycerine into the space between the solid plate and the membrane at a fixed rate, causing the membrane to bulge through the hole and ultimately to rupture. The maximum pressure applied before rupture is defined as the Mullen Burst Test Strength. For various membranes its magnitude has been reported to range between about 25 and 130 psi.

Since the method of sealing to prevent leakage around the periphery of a membrane in an electrodialysis stack is to clamp the edges of the membrane between flat surfaces, it is also important that the membrane be sufficiently strong to resist the clamping forces without excessive deformation or plastic flow. A test for this characteristic has been devised and involves the placement of a stack of 2 in. square membrane samples under a dead load of 150 g/cm^2 for periods up to 7 days, and measuring the change in thickness. This test has proved effective for practical purposes.

iv. Chemical properties: The ability of a membrane to remain stable under continued exposure to saline waters of concentration equal to that of discharged brine, containing widely different combinations of ions, and having pH values varying from acidic to alkaline, is of great importance. This chemical stability is usually

TABLE 13-2

Properties of Some Electrodialysis Membranes

Number[a]	Ref.	Thickness (cm)	Water content (%)	Exchange capacity (meq/dry gram)	Area resistance (Ω-cm²)	Transport No.	Burst strength (kg/cm²)
I CR-61	[6]	0.058	45-50	2.8	11	—	9.8
I AR-111A	[6]	0.058	40-45	2.0	11	—	9.8
Se CMVIO	[7]	0.025	—	2.0	6-8	0.91	6-7
Se AMTL	[7]	0.021	—	2.4	2-4	0.92	9-10
Ne Cl-2.5T	[8]	0.016	30-40	1.8-2.0	2.7-3.2	0.98	3-4
Ne AV-4T	[8]	0.015	20-30	1.5-2.0	3.0-4.0	0.98	7-8
AMF-C	[9]	0.014-0.015	—	1.5	2-4	—	2.4
AMF-A	[9]	0.014-0.015	—	1.4	3-6	—	2.8
ACI-K	[10]	0.23	35-37	2.5-2.8	2.3-3.0	—	—
ACI-A	[10]	0.23	26-28	1.6-1.9	2.2-2.8	—	—

[a] Note: Membranes designated C or K are cationic membranes and those designated A are anionic membranes. Manufacturers of the above membranes are: I, Ionics Inc., United States; Se, Selemion membranes, Asahi Glass Co., Japan; Ne, Neosepta membranes, Tokuyama Soda Co., Japan; AMF, American Machine and Foundry Co., United States; ACI, Asahi Chemical Industry, Japan. There are several other American and foreign manufacturers of electrodialysis membranes including Rohm & Haas, Ionac Chemical Co., and Permutut Co. However, no information on the properties of their membranes was found in the periodic literature.

tested for periods up to 14 days by exposure to 1.0 N acids and bases and to water saturated with chlorine gas.

Another chemical property of fundamental importance is the amount of the membrane which is represented by the exchange sites within the plastic matrix. This is usually expressed in terms of the milliequivalents (meq) of ion-exchange capacity per gram of dry membrane. For different commercial membranes this quantity has been reported to vary from less than 1.0 to about 2.7 meq/g. The higher its value for a given thickness of membrane, the greater should be the capacity of the membrane to transmit ions.

v. Interrelation of membrane properties: While the above properties have been treated separately, they are interdependent. For example, the realization of very high selectivity in a membrane is accomplished by a very "tight" structure, which, as a consequence, causes the electrical resistance to be high and the transmissivity of the ions relatively low. Similarly, if the membrane is designed to have high mechanical strength, the electrical resistance is also increased, thereby involving increased ohmic losses. Thus, most electrodialysis membranes represent compromises of their respective desirable properties. Table 13-2 gives the properties of some of the more common electrodialysis membranes. (See also Refs. [6-10].)

b. Manufacture of Electrodialysis Membranes. As has been stated previously, the first electrodialysis membranes were made by adding ground-up ion-exchange resins to a plastic binder, and using the mixture for forming sheet membranes. Since the ion-exchange resin did not dissolve in the plastic binder, the result was a heterogeneous membrane with charged sites more or less evenly spaced throughout the solidified binder. These early membranes proved unsatisfactory because they lacked dimensional stability and tended to rupture due to absorption of water. The observation that synthetic ion-exchange resins could be produced by polymerizing techniques led to the successful production of homogeneous plastic sheet materials with charged sites of a given electrical polarity. A rather complete discussion of the several processes used in producing electrodialysis membranes has been presented by Wilson [4]. A few summary comments will serve the purpose of this chapter.

The actual methods of producing electrodialysis membranes are proprietary and are described in a multiplicity of patents which have been issued. In general terms there are two classes of procedures followed. The first procedure is to form the membrane from a monomer which contains ion-exchange groups, and after polymerization will have these groups rather uniformly distributed throughout the material. The second procedure is to begin with a polymerized film and to modify the structure chemically so as to provide the desirable ion-exchange groups. The second of these two procedures has been the more successful and is used by several manufacturers. The basic polymeric structures used have included a wide variety of materials, such as a copolymer of styrene and divinyl benzene, a copolymer of polyethylene and polystyrene, a polymer of phenol-formaldehyde, a copolymer of styrene and butadiene, and cellophane. These structures have been modified chemically by processes such as sulfonation for cation membranes, and chloromethylation followed by amination for anion membranes. Some of these membrane materials have been produced in continuous sheets of widths up to 44 in., while others have been made by casting between parallel glass plates. The largest size made by the latter method is 18 by 40 in. Some membranes are reinforced by a fabric backing or insert which adds mechanical strength to the membrane, but also increases the electrical resistance. Materials used for reinforcement include fabrics made from Dynel or glass fibers.

B. Electrodes

The distribution of the electric current over the active area of the electrodialysis cell is accomplished by electrodes at the ends of the stack, constructed of flat metal or carbon plates of the same size as the active area of the membrane. These electrodes are in contact with only that percentage of water which flows through the two end compartments of the stack beyond the membrane pairs. Since the transmission of electricity through the metal electrodes uses electrons as carriers and the solution in contact with those electrodes uses ions as carriers, oxidation and reduction must take place at the electrode surfaces in order to transfer the electricity from ions to electrons and vice versa. Oxidation takes place at the anodic electrode and reduction at the cathodic electrode in order to supply and remove electrons to and from the metallic external circuit, respectively. A disproportionately greater percentage of energy is required for these "plating out" reactions than for the simple movement of the ions through the membranes.

Oxidation at the anode can be the source of erosion of the metal of an electrode in a relatively short time; therefore, it is important to consider possible means of avoiding such erosion. The chemical reactions which occur at the electrodes depend upon the composition and the pH of the solution with which they come into contact. The feed to the electrode compartment, referred to as the "rinse" stream, is generally kept separate from the product and concentrate streams. It is common practice to add acid to the rinse streams in order to prevent alkaline scale deposition in the anodic electrode compartment. Waters which contain chlorides and are acidified to a pH of about 3 undergo oxidation at the anode according to the equation:

$$Cl^- = \tfrac{1}{2}Cl_2(\text{gas}) + e^-; \qquad E_0 = 1.36\ \text{V} \tag{1}$$

where e^- is the electron and E_0 is the electrode voltage. If the rinse water is low in chloride but is acid and contains a large amount of other negative ions, the anode reaction is likely to involve the oxidation of the oxygen from the water molecules according to the equation:

$$H_2O = \tfrac{1}{2}O_2(\text{gas}) + 2\,e^- + 2\,H^+; \qquad E_0 = 1.23\ \text{V} \tag{2}$$

The gases so formed may either remain in solution to be degassed after the waste brine exit, or be degassed before exit. The former gives less chance that all of the gas will be removed from the system.

The above reactions presume that the anode itself is inert. If such is not the case, the anode material will be oxidized and add metallic ions to the solution. Various materials have been used for anodes. Sheets of tantalum, plated with 70 μm thickness of platinum or of titanium plated with platinum have been found to be the most inert. Because of the great expense of these materials, some manufacturers prefer to use carbon blocks, and accept a surface erosion rate of as much as 1 in. per year. Oxidation of the carbon ordinarily results in the generation of carbon monoxide gas as well as the generation of the chlorine and oxygen gases noted above.

The problem in the cathode compartment is much less serious than that in the anode compartment, since the reduction reaction tends to prevent disintegration of the metallic cathode. However, the cations found in natural waters have electrode potentials so high that the water tends to break down and liberate hydrogen gas. Since this would also produce hydroxide ions and aggravate scaling tendencies, the cathode rinse stream is generally acidified to a pH of about 1.8 to 3. This supplies sufficient hydrogen ion for reduction to hydrogen gas without requiring the electrolytic decomposition of water. The material used for cathodes varies, but most typically is stainless steel.

C. End Plates and Stack Alignment

The assemblage of membranes, gaskets, spacers, and electrodes requires compression to prevent leakage around the edges of the several components, and good alignment to minimize fluid friction in the manifolds. To accomplish these objectives, steel end plates are designed to maintain the membranes' flatness as they are pulled together by tie rods or hydraulic presses with sufficient force to reduce leakage to zero or some small permissible amount. Alignment of the membranes and gaskets during assembly can be achieved by punching semicircular notches into the side edges of the components, thus allowing them to slide over temporary alignment rods screwed into end plates, the latter rods being removed after assembly is complete. Holes in the end plates register with the manifold holes in the membrane-gasket assembly and contain fittings for connection to the external piping.

III. PERFORMANCE OF ELECTRODIALYSIS STACKS

The effective performance of an electrodialysis stack can be measured in terms of the current efficiency, concentration polarization, the electrical resistance, and the voltage requirements characteristic of the cell and the stack design.

A. Current Efficiency

The current efficiency, also referred to as the coulomb efficiency, is the ratio of the number of chemical equivalents of salt ions removed from the product stream to the number of Faradays of electricity used. This can be expressed in symbols as follows:

$$\eta_C = [G(C_1 - C_2)F]/(I \times N) \tag{3}$$

where η_C is current or coulomb efficiency, G is flow rate of product water in kg/sec, C_1 is concentration of incoming product stream, g-equiv/kg, C_2 is concentration of outgoing product stream g-equiv/kg, I is current in A, F is Faraday's constant (96, 500 C/g-equiv), and N is number of membrane pairs. Values reported for current efficiencies have been as high as 91 to 96%, although some plants have been reported to have operated with current efficiencies as low as 70%.

Assuming the membrane to be free of pinholes and cracks, and to be adequately sealed to prevent leakage between the product and concentrate streams, the current efficiency will be reduced below 100% by (1) passage of ions of the wrong electrical sign through the membrane, (2) by osmotic transfer of water through the membranes, and (3) by the conduction of electrical current through the water in the manifolds of the stack. The passage of the wrong ions through the membrane will be a function of the permselectivity of the two membranes and has been found to be quite small for most practical membranes.

Water transfer by osmosis is due to the osmotic pressure difference of the solutions adjacent to the two sides of the membrane, and will be proportional to the porosity of the membranes relative to water molecules. In contrast to the electrodialysis phenomenon in which ions are removed from one stream to another, causing one to become the product stream and the other more concentrated, osmotic pressure forces water from the less concentrated (product stream) to the more concentrated stream, thus negating to a small extent the desalting effect of the electric current. This, however, is also a minor factor in the overall process and may amount to about 0.15 ml of water per meq of ions tranferred.

Short-circuiting of the current by conduction through the manifolds is a function of the relative areas of the manifold cross section and the active membrane area. The ratio of manifold area to active area should be of the order of 2% at the most. Of the four manifold streams, the concentrate outlet stream will have the greatest conductivity and the product outlet stream will have the smallest. The two inlet streams will have intermediate conductivities, equal if both are supplied from the same source without recirculation, and unequal with greater conductivity in the concentrate stream using recirculation. The conductivities of all the manifold streams except the product outlet stream will be greater than that of equal areas through the membranes. In spite of this greater conductivity, the leakage current through the manifolds will be small because of the small cross-sectional area of the manifold passages.

B. Concentration Polarization

Polarization in electrodialysis is the substitution of hydrogen and hydroxide ions produced from electrolytic water splitting in order to provide the requisite electrical current carriers through the membranes whenever there is a deficiency of solute ions in the solution adjacent to the membranes. Its tendency to take place at the surface of the membranes is due to the hydrodynamic characteristic of channel flow, in which there is a central turbulent core of flow bounded by thin viscous boundary layers adjacent to the confining surfaces. Whereas ions are transported rapidly across the turbulent core by eddying motion in that part of the stream, their migration through the boundary layers is very slow and is produced primarily by diffusion. These viscous boundary layers impose a resistance to the passage of ions much greater than that of a layer of like thickness in the turbulent core, and hence increase the likelihood of polarization at the membrane surfaces.

The effect of the concentration gradient at the surface of the membrane is one cause for this greater resistance in the viscous layer, and the consequent tendency for polarization to occur produces a paradoxical situation. The ion population at the surface of the membrane is reduced by the withdrawal of the ions required to carry the electricity through the membrane. This reduced ion population, i.e., concentration, at the membrane surface increases the rate of diffusion of the required ions through the boundary layer from the turbulent core, but also increases the voltage drop across this layer which in turn increases the tendency for water to become ionized by electrolysis. The paradox of the situation is that the decreased concentration of ions at the membrane surface which is needed to overcome the diffusional resistance of the boundary layer is also the condition which favors water splitting, and thereby decreases the need for the transport of the solute ions through the boundary layer. While the tendency toward polarization is present at all current densities, it is only at high current densities that the voltage drop across the boundary layer becomes large enough to actually produce water splitting and the consequent substitution of hydrogen and hydroxide ions for the solute ions as current carriers.

Since the tendency for polarization increases with current density, there will be some critical current density above which polarization is relatively certain to occur. Because of the complicated nature of the flow phenomenon and its dependence upon many variables, including the salinity and temperature of the product water stream, the geometrical configuration of the flow path, the velocity and degree of turbulence of the flowing water, and the spacing of the membranes, it is too complex to estimate this critical current density from basic data; it must be determined experimentally. Experimental tests must be made for each given design and size of cell, the results of which are the basis for a given plant design.

Although the critical current density cannot be calculated accurately from basic data, there have been devised certain indicators that suggest approximate limits. One such method for determining the critical current has been explained by Cowan [5] in which the ratio of voltage to current, V/I, is plotted against the reciprocal of current. On the basis of numerous test data in which voltage was measured in electrodialysis cells for varying values of current, he verified the fact that his test measurements, when plotted as the ratio of voltage to current against the reciprocal of current, fell on a very indicative curve, decreasing linearly at the right-hand end of the curve (low current density) as the current increased. At some critical value of current the slope of the line decreases, passes through a zero value, and then increases in a direction opposite to that for the lower values of current. A schematic drawing of such a plot is shown in Fig. 13-4. In seeking an explanation for the rapid increase of V/I after the critical low point was reached, he studied the salinities and the pH of both the product and concentrate streams, and observed that the pH of the outgoing product stream decreased and that of the concentrate stream increased when the current was greater than the critical current. The explanation for this is probably related to the water splitting that occurs at currents greater than the critical current. Polarization apparently takes place more actively at the anionic membrane; hence, the OH^- ions migrate through the membrane to the concentrate compartment, leaving H^+ ions in the viscous surface layer in the product compartment. From a practical standpoint, if polarization is to be avoided, the maximum operating current should be well below the critical value, even as low as 70% of this value being recommended in some cases.

Thus it can be deduced that polarization may well be a significant explanation for the excessive resistance associated with high current densities, and therefore it becomes an important reason for limiting the current density to be used in any given situation. On the other hand, the above need to limit current density in order to control polarization has been shown to have an undesirable effect on the basic economy of the electrodialysis process in that the less the current density, the greater the need for large membrane area, and hence greater costs. Polarization imposes a limit on current density, and it also results in limiting the economy of membrane costs.

One proposal for both decreasing membrane area and increasing current density, but doing so in a manner to better control polarization, has been the suggestion of a "tapered" stack. In such a device the membrane and spacers would have a greater width at the entrance than at the exit, and the concentration of the saline water feed at the

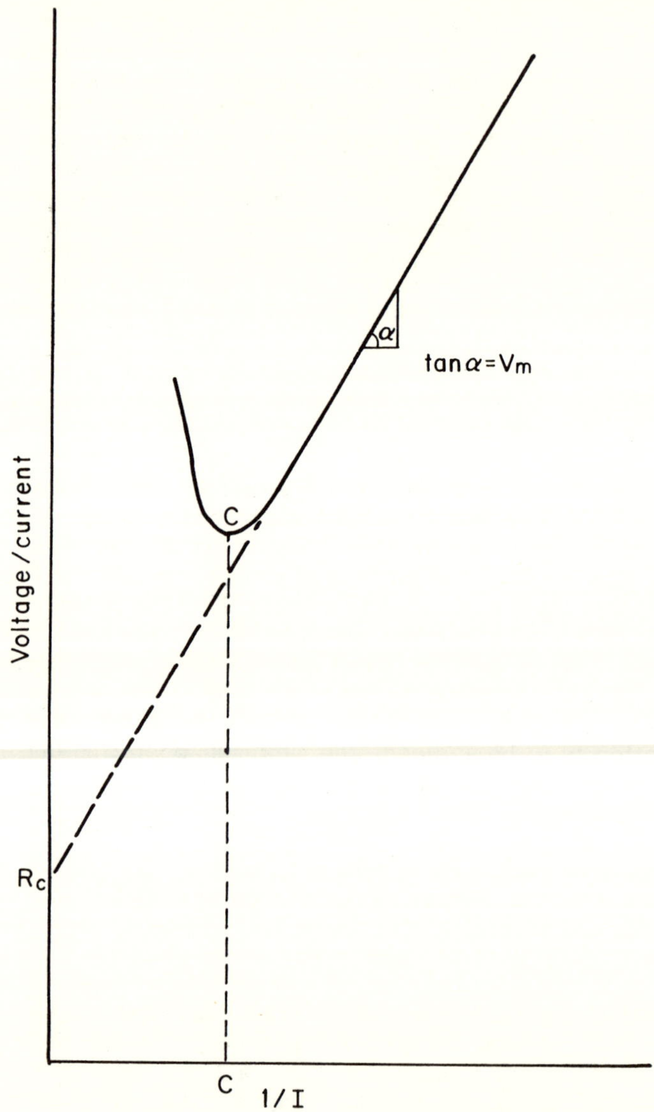

FIG. 13-4. Plot to determine the limiting current density, after Cowan [5] (schematic only). Limiting current density (C) corresponds to the minimum point on the curve, indicated by the broken line, C--C.

entrance would be great enough to preclude polarization. The increased water velocity along the narrowing flow path toward the tapered outlet and the accompanying turbulence would bring more solute ions into contact with the membrane surface than would be possible in a rectangular stack of the same membrane area, and would thus reduce the tendency toward polarization at the tapered outlet, while at the same time permitting the use of greater current density. Carried to extreme, a sufficient number of "tapered" stacks, placed side by side with the wide ends outermost, would form a circular ring-shaped stack. If large circular membranes were available the edge sealing between segments would be eliminated, and the increased turbulence of the flow path would increase the effectiveness of the area utilized. Although there has been some experimentation using "tapered" stacks, circular membranes large enough to realize the above conditions have not as yet been produced. Therefore, this concept has not been utilized in a practical stack, but does illustrate the possible use of design innovations to overcome the problem of polarization.

Polarization is objectionable not only from the standpoint of the inefficient increase in energy consumption, but also because the change in pH of the concentrate stream tends to cause scale deposition. Hydroxide ions emerging from a membrane surface into the concentrate stream will tend to combine with magnesium ions, if such are present, forming the relatively insoluble magnesium hydroxide scale. Spiegler [11] and Weiner [12] treat polarization more fully.

C. Cell Resistance

The basic electrical resistance of an electrodialysis cell is the sum of the resistances in the two membranes and the two water streams in the product and brine compartments, and has been determined to be a fairly constant value. The resistance of the membranes depends on the membrane material and thickness and on the ions which pass through them. As already indicated, the measure for this resistance is customarily determined by application of an ac voltage with a frequency of 1000 Hz. The value so obtained is useful in comparing one membrane with another, but actually is somewhat less than the value which applies during normal operation using dc current. Expressed in terms of area resistance, i.e., the ratio of voltage drop to current density, the values for resistances that have been listed by various manufacturers for different membranes vary from about 2 to 20 Ω-cm^2.

The resistances of the water streams in the brine and product compartments are dependent upon the concentrations of the streams and vary from point to point along the flow path. The resistance in the brine stream is reduced to quite a low value because of recirculation, and is relatively unimportant in the total resistance as compared with the increasingly high resistances in the product stream. The resistance of the product stream is particularly important because of its high value; therefore, any change in temperature of the water and spacing between the membranes in this stream produce further variations that are significant in the total resistance.

The extreme and average values existing in the product and concentrate streams can be calculated from data on the electrical conductance of the particular saline water being analyzed. For example, the design characteristics of the Ionics Mark III stack, described by Katz [6], states that a current of 100 A is capable of demineralizing by 46% a 3000-ppm saline water feed at the rate of 250,000 gpd. The outlet product water from the stack would therefore have a salinity of 1620 ppm. Assuming the solute to be NaCl and the stream width to be 0.1 cm, the corresponding area resistance of the product stream would be 17.6 Ω-cm^2 at entrance and 31.5 Ω-cm^2 at exit, based on equivalent conductance data for NaCl solutions at 25°C. To estimate the resistance of the concentrate stream it is assumed that this stream flows at the same rate as the product stream. However, in order to more efficiently conserve the moderately brackish water for the product stream, the outlet water from the concentrate stream is mixed with a small amount of the brackish feedwater and recirculated in the concentrate stream, thus raising its salinity to a value very much greater than the saline source. When this is done, salinity values as high as 13,620 ppm at entrance and 15,000 ppm at exit have been reported to obtain, having corresponding resistance values of 4.25 and 3.89 Ω-cm^2, respectively.

As mentioned in the first paragraph, the membranes represent a component of the total cell resistance as well as the water streams. The resistance of membranes fluctuates with the concentration of the solution with which they are in contact, and the variations of this phenomenon have been studied by Katz [6]. He has plotted curves relating membrane resistances to concentration, and from these curves the resistance for both the positive and negative membranes at the above concentrations each appear to be 13 Ω-cm^2. The total cell resistance, therefore, is the sum of these two membrane resistances plus the average resistances of the two water streams, totaling approximately 54 Ω-cm^2.

In the above example the conductance of NaCl solution was that corresponding to a temperature of 25°C, because most electrodialysis desalters are operated at approximately ambient temperature. However, temperatures higher than 25°C would be advantageous, since the conductance increases rapidly with temperature, and an increased conductance produces less ohmic power loss. For example, the equivalent conductance of NaCl solutions increases almost linearly with temperature, the rate of increase varying with concentration and being greatest for very dilute solutions. If the cell in the above paragraph could be operated at 90° rather than at 25°C, the equivalent conductance would be increased sufficiently to reduce the average resistances of the product and concentrate streams to about 9 and 1.5 $\Omega\text{-cm}^2$, respectively. If the membrane resistances did not change, the cell resistance would be reduced from 54 to 36.5 $\Omega\text{-cm}^2$, which would reduce the ohmic losses by over 30%. The attractiveness of this possibility has led to some experimentation with high-temperature operation of electrodialysis desalters, and there may be production plants using elevated temperatures in the not too distant future.

D. Voltage Requirement

As mentioned in the previous section, the resistance in an electrodialysis cell is a composite resistance, depending on the voltage required to produce a given current in the cell. This required voltage is the sum of three components, i.e., (1) the ohmic voltage drop through the cell, IR_c, using R_c from the previous paragraphs (2) the membrane potential, V_m, due to the difference in concentrations of the liquids on the two sides of the membrane; and (3) the polarization potential. The value for V_m is a fairly constant value for different values of current, I. The polarization potential, V_p, has significance only when the cell is operating under polarized conditions in which water splitting occurs. The sum of these components can be written algebraically as follows:

$$V_c = IR_c + V_m + V_p \tag{4}$$

When the cell is operating under nonpolarizing conditions, $V_p = 0$ so that Eq. (4) becomes:

$$V_c = IR_c + V_m \tag{5}$$

or, rewritten in the following form:

$$V_c/I = R_c + V_m/I \tag{6}$$

When experimental data are applied to Eq. (6) the values of R_c and V_m have been demonstrated to be very nearly constant in practical electrodialysis cells operating at a given salinity. As a result, when V_c/I of Eq. (6) is plotted against $1/I$, as in Fig. 13-4, the curve in the right-hand region of low current values, I, where V_p is not a factor, is a straight line with a slope of V_m intercepting the vertical axis of $1/I = 0$ at R_c. Figure 13-4 is a schematic curve only, but is similar to many experimental curves having the same coordinates, and on which experimental data are plotted showing numerous points falling very close to a sloping straight line for electrical current values less than those which cause polarization. Cowan's [5] article further illustrates the complicated effect of V_p and pH in their relation to resistance when currents greater than the critical current are applied to an electrodialysis cell.

IV. MODIFIED ELECTRODIALYSIS DESIGNS

Because polarization in an conventional electrodialysis stack presents the problem of scale deposition on the anionic membranes due to passage of hydroxide through those membranes, there have been several proposals to eliminate the anionic membrane from the stack. Two of these proposals, known as electrogravitation and transport depletion, have received sufficient attention to be mentioned here. In electrogravitation the anionic membrane is completely eliminated and the stack consists of cationic membranes only. In transport depletion the anodic membrane is replaced with a neutral membrane, thus forming pairs of cationic and neutral membranes.

Electrogravitation, as the name implies, involves the use of gravity as a means of separating the dilute and concentrate streams. In this scheme the cationic membranes are placed vertically and spaced somewhat farther apart than those in conventional electrodialysis, with the feed entering at the bottom of the stack. Because of the selective nature of the cationic membrane, there will be a relatively dilute boundary layer on one side of it and a concentrated boundary layer on the other side. Since the concentrate is more dense than the dilute solution, the concentrate will tend to flow downward, and the dilute, or product, will tend to rise. If the liquid from the dilute side of the membranes is withdrawn from the top of the assembly and that from the concentrate side is withdrawn from the bottom, it is assumed there will be the desired separation of product water from concentrate. Apparatus to utilize this principle could be conceived of as consisting of an open tank in which the membrane sheets are suspended vertically.

Saline water could be introduced at a rate such that the level in the tank would remain constant, and the rates of withdrawal of product water and brine would be such that they would not interfere with gravitational convective currents in the spaces between the membranes. Although this idea has been the source of considerable interest and some research in experimental laboratories, no production plants of this type have come to the attention of the writer.

The transport depletion method of desalination is very similar to normal electrodialysis, in that the stack consists of pairs of membranes, alternating cationic and neutral. Spacers and other cell components are the same as in the normal apparatus. When an electrical current is caused to flow through the stack, ions of both electrical signs pass through the neutral membrane, so that there is no depletion or concentration in the boundary layers on its two sides. If polarization occurs at the cationic membrane no harm is done, since the hydrogen ions formed by water splitting go into the concentrate compartment and tend to dissolve any alkaline scales which may be present. The hydroxide ions remain in the product compartment where the ionic concentration is too low to form scale. As yet little laboratory work and no production plants have been undertaken because the apparatus costs would be equal to conventional electrodialysis, and the current efficiency would be considerably less because of the lack of selectivity of the neutral membranes.

Another modification of the conventional design that has met with questioning interest is one which attempts to reduce the power costs of a normal electrodialysis operation by utilizing the electrical potential due to differences in solution concentration as the driving potential in an electrodialysis cell. One version of an apparatus for accomplishing this purpose proposes an arrangement of cell pairs similar to a conventional stack, but with seawater or some concentrated solution fed to every fourth compartment and less concentrated saline water to the three intermediate compartments. A "salt bridge" would be connected externally to complete the circuit, so that a concentration potential will result in a current through the membranes. The first and thrid would be the concentrate compartments and the second and fourth, the product. Although there has been considerable laboratory work to research this idea, the present development indicates that the costs of desalting water by this process are even greater than the costs of normal electrodialysis, so that as of this date this scheme has not reached the production plant stage.

V. DESIGN CALCULATIONS

The design of an electrodialysis desalting plant to produce a given quantity of water of specified quality from a source of known salinity is generally done by the manufacturer. The designer in such a case will have available the proprietary information concerning performance of the particular membranes and other components basic to the selection of the number of stages required, the number of stacks per stage, the number of cells per stack, the dc voltage and current needed, and the fraction of saline water to be recovered as desalted water. Since a detailed design is beyond the scope of this book, the published ratings by Katz [6] for the Ionics Mark III Membrane Stack will be presented to illustrate some of the design calculations which need to be made. These data, including some items previously cited, are as follows:

Maximum hydraulic flow rate	250, 000 gpd
Pressure drop at maximum flow	60 psi
Number of membranes	550
Size of membranes	18 × 40 in.
Total area of membranes	2750 ft^2
Percentage of total area available for ion transfer	70%
Demineralization (70°F Cl water, full flow)	46%
Current required for 3000 ppm feed	100 A
Voltage required for 3000 ppm feed	1000 V
dc power for 3000 ppm feed	100 kW
dc energy consumption for 3000 ppm feed	9.5 kWh/kgal

Using these values, the following calculations are of significant interest.

A. Electrical Current Density, i

$$i = \frac{I}{\text{area}} = \frac{100}{(18 \times 40 \times 2.54^2 \times 0.70)} = 30.7 \ \frac{\text{mA}}{\text{cm}^2}$$

This value, arrived at by using the above experimental data, represents the electrical current density that probably has been chosen by the designer for operating the plant close to but less than the critical density at which polarization is likely to occur. One criterion commonly used for determining the critical current density is i/C_p, where C_p is the concentration of the product water stream. The value of i/C_p is dependent upon the cell design, water velocity, and water analysis, etc. By using the figures of the above Ionics Mark III plant, and assuming NaCl to be the solute and the percentage of demineralization to be 46%, the value of C_p at inlet (3000 ppm) is calculated to be 51.4×10^{-3} g-equiv/kg and 27.8×10^{-3} g-equiv/kg at exit (1620 ppm). Based on the above computed value where i = 30.7 mA/cm^2, the corresponding ratios of i/C_p for the above concentrations will vary from 0.597 to 1.10 A $\times$ kg/(g-equiv $\times$ cm^2), the average being 0.850. These values can be considered as representing acceptable ratios in the above plant and will have significance for designers who must determine an acceptable i/C_p ratio for any different prescribed set of conditions.

The most advantageous value for current density is dependent on economic factors as well as upon the problem of polarization. The cost of desalting water with an electrodialysis device includes some items whose costs are proportional to the current density, such as the electricity and the amortization of the rectifier and control equipment, and some items for which costs are inversely proportional to the current density, including amortization and replacement of the membranes, spacers, electrodes, and other stack components. Under these conditions, there will be some current density for which the sum of these two categories of cost is a minimum, i.e., the current density for minimum product cost. It would be desirable for this current density to coincide with that for avoiding polarization. This could be brought about during the initial cell design by varying stream velocities, spacer thickness, etc., until the above conditions are realized. Whether or not this was done in the case of the stack under consideration is not known, but it is reasonable to expect that this widely used design of stack is correctly proportioned and operated to give at least near-maximum economy with the operating current density very close to the critical current density for polarization.

B. Current Efficiency, η_C

$$\eta_C = [G(C_1 - C_2)F]/(I \times N)$$

where G is flow rate (kg/sec), C is concentration (g-equiv/kg), F is Faraday's constant (C/g-equiv), I is current (A), and N is number of pairs of membranes.

$$\eta_C = \frac{250,000 \times 8.33}{24 \times 3600} \times \frac{(51.4 - 27.8)10^{-3} \times 96,500}{100 \times 275} = 91.4\%$$

This current efficiency probably would have been known to the designer from previous experience, and could have been used by him to either calculate the current needed for a given amount of demineralization, or to estimate the amount of demineralization to be accomplished by a given electrical current.

C. Residence Time of Water in the Stack, $\Delta\theta$

$$\Delta\theta = \frac{\text{P. L.}}{v} = (C_1 - C_2)\rho\left(\frac{Fd}{i\eta_C}\right) = \text{time in seconds}$$

where P. L. is path length through stack (cm), v is water velocity (cm/sec), ρ is density of water (g/cm^3), and d is spacer thickness (cm). For the conditions assumed, the time of residence is as follows:

$$\Delta\theta = (51.4 - 27.8) \times 10^{-6} \times \frac{1.0 \times 96,500 \times 0.10}{30.7 \times 10^{-3} \times 0.914} = 8.11 \text{ sec}$$

The above equation was derived by equating, for the same interval of time, the coulombs available from the current to the coulombs accounted for in the desalting of the water. The value derived for this residence time can be used by the designer to check the velocity of flow through the stack with respect to the path length. The path length in the Mark III stack is not exactly specified in the reference, but is said to be long, between 10 and 15 ft. Using the larger of these two values, the magnitude of the water velocity would be 56.4 cm/sec.

VI. COST OF DESALTED WATER

Four major cost elements in the production of desalted water by electrodialysis are capital amortization, membrane costs, operation costs, and maintenance costs other than membrane replacement. Although these elements will vary with the size of plant and many other variables, it is instructive to consider the published figures relating to the 650,000-gpd plant at Buckeye, Arizona as given by Katz [6]. In this plant the salinity is reduced from 2100 to 400 ppm by installing

TABLE 13-3

Annual and Unit costs of Desalted Water
for a 650,000-gpd Plant

	48% capacity		98% capacity	
	($/yr)	(¢/kgal)	($/yr)	(¢/kgal)
Bond amortization	20,500	18.3	20,500	8.8
Membrane replacement	9,900	8.8	9,900	4.2
Maintenance	7,800	7.0	7,800	3.4
Operation	18,780	16.8	38,720	16.6
Total	56,980	50.9	76,920	33.0

two stages of three Mark III Ionics stacks per stage (these stacks are described on page 296). The cost elements of this plant [6] are summarized in Table 13-3.

The figures in Table 13-3 are representative of the present state of development of electrodialysis. Since the current efficiency of the Mark III stack, as estimated on page 296, is over 90%, reductions of unit water cost below the 33¢/kgal given in the last column will require decreases in capital cost and in membrane replacement cost. Such decreases could result from the development of membranes that either cost less or last longer than the present ones. Research and development efforts are directed toward these ends.

Table 13-3 also shows the effect of operating the plant at partial capacity, indicating that it is the operating costs that are chiefly affected by percentage capacity, thus making the contributions of the first three elements of cost increasingly important as percentage capacity is decreased below 100%.

Another approach to the reduction of unit water cost is optimization with respect to current density. As is pointed out by Lacey et al. [13], an increase in current density in a plant of given capacity is concordant with a decrease of the required membrane area and a consequent increase in the voltage drop through the stack, thereby producing a need for an increase in electrical power consumption. The corresponding changes in the several cost elements can be illustrated by plotting total annual cost versus current density, the curve of which is U shaped. Lacey points out that the current density corresponding to the minimum point of this curve, i.e., where the annual cost is lowest, may be greater than the limiting current for polarization, and hence impractical. Nevertheless, it is important to have this basic understanding of the conditions under which optimum operational costs could be realized.

REFERENCES

1. W. Juda and W.A. McRae, U.S. Patent Nos. 2,636,851 and 2,636,852, 1953.
2. E. Wegelin, Electrodialysis, Proc. Saline Water Conversion Symposium, Washington D.C., 1957.
3. Bureau of Reclamation: Test Manual for Permselective Membranes. Res. Develop. Progr. Rept. No. 77, Office of Saline Water, U.S. Dept. Interior, 1964.
4. J.R. Wilson, Demineralization by Electrodialysis, Butterworths, London, 1960.
5. D.A. Cowan, Interaction of Technical and Economic Demands in the Development of Large Scale Electrodialysis Demineralizers, Advan. Chem. Ser. 27, 1960, pp. 224-231.
6. W.E. Katz, Some Practical 1959 Advances in Electric Membrane Demineralization, Advan. Chem. Ser. 27, 1960, pp. 232-246.
7. S. Itoi, Ionic Membrane Electrodialyser for Marine Use, Desalination, 2 (No. 3), 1967, pp. 378-386.
8. T. Matsuda, S. Ogawa, and Y. Onoue, Demineralization of Sea Water by Electrodialysis, Desalination, 3(No. 3), 1967, pp. 295-307.
9. W.K.W. Chen, M.S. Mintz, and D.G. Conning, Electrodialysis and Its Application to a small Household Desalting Appliance, Dechema Monogr. 47, pp. 619-637.
10. M. Seko, Large Scale Brackish Water Conversion—Electrodialysis Demonstration Plant at Webster, South Dakota, USA, Dechema Monogr. 47, pp. 575-600.
11. K.S. Spiegler, Polarization and Water Splitting in Electrodialysis, Notes for Extension Short Course on Desalination, 1969, University of California, Berkeley, 20 pages, 1969.
12. S.A. Weiner, P.M. Rapier, and W.K. Baker, Polarization Characteristics of Electrodialytic Demineralizers, I & E.C. Process Design and Development, 3, 1964, pp. 126-129.
13. R.E. Lacey, E.W. Lang, and E.L. Huffman, Economics of Demineralization by Electrodialysis, Advan. Chem. Ser., 38, 1963, pp. 168-178.

Chapter 14

REVERSE OSMOSIS

Reverse osmosis, as a process for water desalination, is the newest of the commercial processes. Even in its present early stage of development it is more economical than distillation when used for waters with low salinities, and has promise of being successful with even seawater. It is competitive with electrodialysis, but is in a much earlier stage of development than the latter. Millions of dollars are being spent on improving the equipment of this process, with the result that there are continually many new and increasingly large installations being built. Several companies have already marketed small to moderately sized units, and several moderately sized units have been used as experimental pilot plants, some of which are listed in Table 14-1.

The essential successful feature of the process, the membrane, was first described by Loeb et al. [1] in terms of a procedure for casting such membranes from a solution of cellulose acetate. The research and development which followed is described by Lacey and Loeb [2] and by Merten [3]. These efforts are continuing, with new discoveries giving promise of more effective membranes in the future. While the basic principles of reverse osmosis have been outlined in Chapter 4, it is essential to review these principles in order to logically proceed with the application of reverse osmosis to particular pieces of apparatus.

TABLE 14-1

Partial List of Reverse Osmosis Desalination Plants

Location	Capacity 1000 gpd	Type	Year	Water salinity (ppm)
Coalinga, California	5	1 in. Diameter tubular	1965	2475
Bari, Italy	10	Spiral-wound 4 in. tube	1966	5000
San Diego, California	50	Spiral-wound 4 in. tube	1967 1968	4500
Bessie, Oklahoma	10	Spiral-wound 4 in. tube	1968	3200
Midland, Texas	7	Spiral-wound 4 in. tube	1968	1372
Portable Test Unit	50	Plate-and-frame-36 in. diameter	1968	Various
Plains, Texas	100	Hollow fibers	1969	—
Fort Stockton, Texas	2500	Hollow fibers	1970	—
San Diego, California	20	Porous tube	1970	1500-4500
Test unit	87	Tubular	1970	Various
Test unit	32	Spiral-wound	1970	Various
Greenfield, Iowa	150	Hollow fibers	1970	2300
Manassas, Virginia	270	Hollow fibers	1971	—

I. THE OSMOTIC PHENOMENA

Osmosis is a term indicating the natural passage of pure water through a membrane without the use of external force. It has been studied for more than 200 years in connection with plant and animal membranes, where it was observed that pure water passage through membranes separating two solutions of different concentrations took place in a direction that tended to equalize the concentrations of the liquids on the two sides of the membrane, i.e., the water component moved through the membrane from the less concentrated solution to the more concentrated one. This was observed to be a natural phenomenon requiring no external force other than what has come to be known as the "osmotic pressure" of the liquids of the two different concentrations. The significance of this concept in explaining many organic processes in plants and animals led to the idea that such a phenomenon could be applied to desalination.

Many attempts to quantitatively measure the movement of water through plant and animal membranes were studied, but were not very fruitful due to their extreme variability. Attention was turned, therefore, to the fabrication of reproducible synthetic membranes for systematically studying the movement of water under controlled conditions. The history of such synthetic membranes, used for many years to measure osmotic pressures of various solutions, has been reviewed by Findlay [4]. Conclusions derived from his tests and those of many others have greatly increased the understanding of osmosis and have aided in its application to the desalination of water.

The classical demonstration of osmosis involves fixed quantities of two solutions of different concentration separated by a membrane which is permeable to water but impermeable to dissolved salts. The more concentrated solution is contained in a closed vessel which is fitted with a pressure gauge. The less concentrated solution, generally pure water, is contained in a vessel open to the atmosphere. At zero time the levels of the two liquids are equal, with the pressure gauge on the closed vessel indicating 1 atm. As time passes the level of the pure water falls at a decreasing rate while that in the closed vessel rises correspondingly, thereby increasing the pressure on the gauge. In such an experiment the pressure rises to a specific critical value, referred to as the "osmotic pressure," after which the flow of water through the membrane ceases. If a source of external pressure, greater than the osmotic pressure, is now applied to the closed vessel, this increased pressure causes the flow of water through the membrane to be reversed, i.e., water will flow out of the more concentrated solution and into the dilute solution in the open vessel. This reverse phenomenon is the basis for the origin of the term "reverse osmosis," by which the process of this concept of desalination is generally designated. The classical demonstration, noted above, assumes membranes which are completely impermeable to the dissolved salts and which can be operated in either direction. Actually, synthetic membranes are not perfectly selective and do permit some ions to pass. They also are unidirectional, so that a membrane must be used with the proper face on the high pressure side to minimize the passage of ions, since when it is reversed it will not exclude many of the ions.

II. MAGNITUDE OF OSMOTIC PRESSURE

Values for the osmotic pressure of ionized solutions cannot be calculated exactly from purely theoretical considerations because of the imperfect state of knowledge concerning the degree of ionization and the completeness of dissociation. The present concept seems to favor the theory that salts of strong electrolytes in solution are completely ionized regardless of the concentration. However, at the same time it must be recognized that there are electrical forces of attraction between component ions which produce the effect of incomplete dissociation, so that the more concentrated solutions "appear" to be less dissociated than do dilute solutions. Only the very dilute solutions perform as though completely dissociated. The implication of this situation, so far as osmotic pressure is concerned, can be appreciated by briefly reviewing the various equations which have been formulated in an attempt to correlate experimental values of osmotic pressure.

One early attempt to relate measured values of osmotic pressure to other properties was based on the idea that dissolved materials in solution are in random motion. Assuming that the random motion of the dissolved particles is analogous to that of the molecules of a Perfect Gas, the pressure due to the moving particles (the osmotic pressure) can be expressed by an equation similar to the Perfect Gas Law, namely:

$$\pi = RTn/v \qquad (1)$$

where π is osmotic pressure, n/v is moles per unit volume, R is the gas constant, and T is temperature. Equation (1) has also been derived from other arguments. It was first deduced from experimental data by the classical work of van't Hoff [5]. More recently it has been obtained by modification of Eq. (3), which follows, using the approximation that the activity of the solvent is equal to its mole fraction. Regardless of the method of derivation, the end result could be shown to have the form of the Perfect Gas Law.

Equation (1) has proved to give results of the correct magnitude for un-ionized solutions of low concentration, such as for sugar; but for ionized saline solutions, such as those of interest in desalination, the values appear to be too low. For example, the osmotic pressure for 1 molal solution of sodium chloride at 20 °C is 352 psi when computed from Eq. (1). This is considerably lower than its measured value of 650 psi. To account for this discrepancy it has been argued that each of the ions must be considered as a separate unit, or, in other words, that one mole of sodium chloride dissolves to produce two "moles" of ions. If such is the reasoning, the osmotic pressure computed from the above equation would thus be doubled, or 704 psi, which then is higher than the measured value. It has been argued that the discrepancy between this latter computed value and the measured value can be attributed to incomplete dissociation of the sodium chloride.

The above discrepancies have led to many investigations of solution properties, one of the most important being that of Debye and Hückel [6], whose mathematical analyses were based on the assumption that dissolved ions of one electrical sign are surrounded by "atmospheres" of ions of the opposite electrical sign. Their findings were general, applying to many facets of solution properties, and have been related to osmotic pressures by other investigators in terms of modifications of Eq. (1). The modified form of the equation, as reported by Robinson and Stokes [7], is as follows:

$$\pi = (RT\nu m M_A/\bar{V}_A 1000)\,\phi \tag{2}$$

where π is osmotic pressure, m is molal concentration, ν is the number of ions per mole, M_A is molecular weight of solvent, $\bar{V}_A$ is partial molal volume of solvent, and ϕ is an osmotic coefficient, obtained from boiling point data for sodium chloride. It will be observed that this equation is similar to Eq. (1) and differs from it chiefly in the inclusion of an empirical coefficient to specify the ionic concentration. This osmotic coefficient, ϕ, has been determined and tabulated for a number of single component solutions. The values for sodium chloride, which are of particular interest in the field of desalination, vary from 0.93 at $m = 0.1$ to 0.989 at $m = 2.0$, this being the pertinent range of concentration of waters to be desalted by this process. Since these values of ϕ are determined from experimental data, the results given by Eq. (2) more closely agree with measured values of osmotic pressure than when computed from the more simple Eq. (1).

The equation that is considered most basic and exact for determining osmotic pressure is one that can be obtained from thermodynamic reasoning (see Reid [8]), written as follows:

$$\pi\bar{V}_1 = RT \ln a_1 \tag{3}$$

where $\bar{V}_1$ is the partial molal volume of the water component and a_1 is the activity of water. Other symbols are as before.

The one idealization involved in Eq. (3) is that the solvent water component is incompressible. The practical difficulty of Eq. (3)

is that values of the activity, a_1 , are not generally available. Because of this, the activity is often replaced by a ratio of vapor pressures, leading to the following form:

$$\pi\bar{V}_1 = RT\ln(p_0/p) \quad (4)$$

where p is vapor pressure over solution and p_0 is the saturated vapor pressure of pure water. It should be noted that the pressure ratio of Eq. (4) is closely related to the boiling point elevation, and so may be obtained from data on that quantity. This approximation is valid for dilute solutions only, which classification includes most of the range of concentrations of interest in reverse osmosis desalination. Values computed from Eq. (4) agree with those from Eq. (2), since both equations involve empirical information derived from experimental data.

The concept of Eq. (4) has been analyzed by several investigators. For example, Millard [9] computed the net work done during a reversible isothermal cycle in which he conceived a mole of solvent to be expelled from a solution by an "osmotic piston." This mole of solvent is vaporized at saturation pressure, expanded to the partial pressure over the solution, and finally condensed into the latter. When the sum of the above work calculations is equated to zero, as required by the second law of thermodynamics, the result is Eq. (4) Besides the assumption that the solvent vapor obeys the Perfect Gas Law, the only other approximation involved in this calculation is that the specific volume of the liquid is negligible when compared with that of the vapor.

Although osmotic coefficients and pressures have been determined for a large number of different solutions, it is chiefly those for sodium chloride and sea salt solutions which are particularly applicable to desalination. The osmotic pressures for solutions of these two materials have been determined and are plotted on Fig. 14-1, in which the osmotic pressure is shown as a function of the molal concentration of the solution, with temperature as a parameter. It has been observed that a single curve represents the osmotic pressures of both seawater and sodium chloride solutions when these coordinates are used. This equality of osmotic pressures probably reflects the fact that over 85% of the mole content of the sea salt is sodium chloride. Also plotted on Fig. 14-1 are values for the percentage salinity as a function of the molal concentration. Here, the curve for sea salt solutions is seen to lie above that for sodium chloride, reflecting the fact that the equivalent molecular weight of the sea salt is greater than that of sodium chloride.

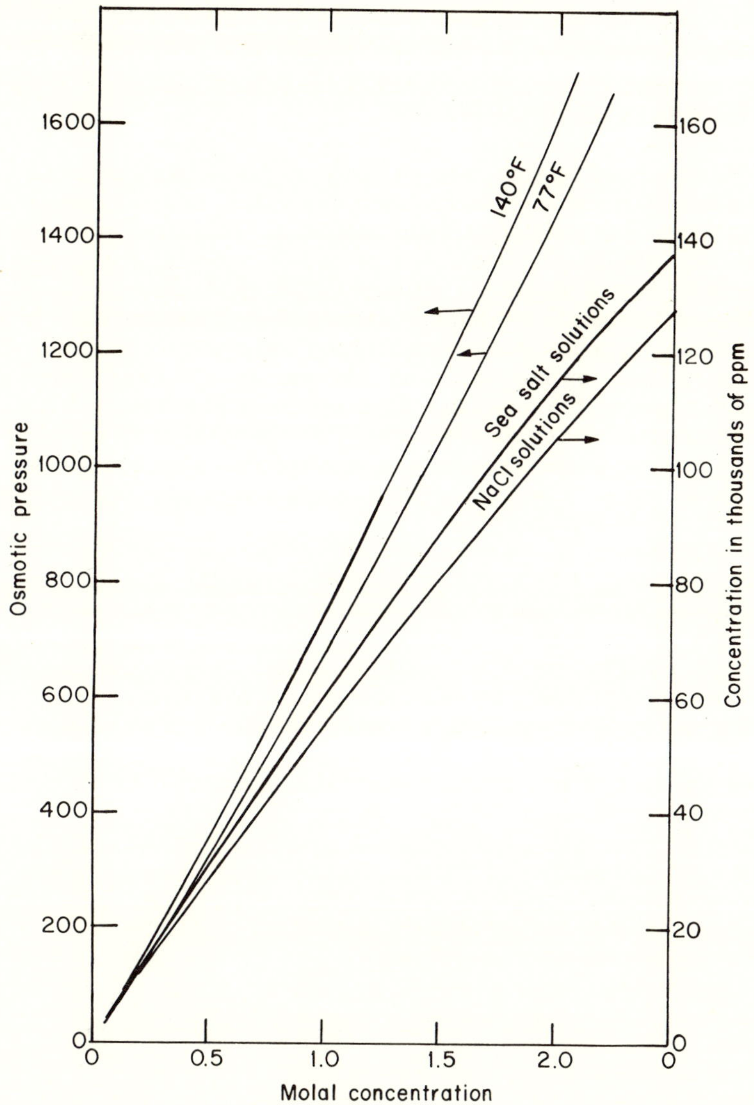

FIG. 14-1. Osmotic pressure for sea salt and sodium chloride solutions. Data for curves obtained from Figs. 11:520 and 12:520, OSW Data Book, Ref. [8], Chap. 2.

From these curves it can be seen that the magnitude of osmotic pressure for a sea salt solution of 0.552 molal concentration, normal seawater, is about 360 psi. For concentrations less than this, the osmotic pressure is proportional to the concentration, while for higher pressures the curve deviates from a straight line in a concave upward direction, thereby increasing the osmotic pressure more than in proportion to the concentration.

Osmotic coefficients are available for many two-component solutions other than sodium chloride and sea salts, and are published in tabular form. For multicomponent solutions, including most saline waters used for desalination, the values of osmotic pressure can be approximated in terms of the principal dissolved material, or in terms of a sea salt solution having the same molal concentration. In any case, minor inaccuracies in determining the osmotic pressure will have little effect on the design of reverse osmosis desalination equipment, since the operating pressure of such a device will be several times the osmotic pressure. For example, it is usual to operate reverse osmosis desalters at pressures of about 600 psi for most saline water, even though the osmotic pressure of such solutions may be less than 100 psi.

In summary, it is possible to approximate the osmotic pressure of saline solutions from existing data with a reasonable degree of accuracy. Osmotic pressure increases with concentration, so that the operating pressure of a desalting device must exceed the osmotic pressure corresponding to the concentration of the waste brine at outlet from the device. Also, osmotic pressure increases with temperature, approximately in proportion to the absolute temperature, so that any rise in temperature must be compensated by an increase in operating pressure.

III. MEMBRANES FOR REVERSE OSMOSIS

A. Membrane Materials and Fabrication

The application of reverse osmosis to the desalination of water is dependent upon the production of membranes which will reject most of the ions and at the same time permit water to pass at an acceptable rate. From the wide variety of osmotic membranes which appear in nature, it can be inferred that there should be many possible materials for producing synthetic membranes possessing osmotic properties. This inference is borne out by experiment. However, only a few have proved attractive for water desalination, the chief of these being variations of cellulose acetate. Reid [8], between 1953 and 1957, found that commercial cellulose acetate films were capable of excluding dissolved ions found in seawater, but that the permeability to flow of pure water through such films was too small to be practical for desalination. Loeb and Sourirajan [1] were able to overcome this problem when they found that cellulose acetate, in the form of cast membranes, could be made to satisfactorily permit passage of large amounts of water by the addition of certain leachable salts to the casting solution. Many basic materials other than cellulose acetate have been tested in membranes, such as nylon, which was used by the Du Pont company in its work with hollow tubes.

In order to appreciate the problem of membrane fabrication it is desirable to review the early procedure and formulation published by Loeb and Sourirajan [1] and to note its later modification. The scheme for fabricating these membranes in sheet form is as follows:

1. Use the following formula to prepare the casting solution and maintain it at a temperature of - 8° to - 11 °C:

Cellulose acetate	22.2% by weight
Magnesium perchlorate	1.1% by weight
Distilled water	10.0% by weight
Acetone	66.7% by weight

2. Prepare a casting plate for the membranes. This is usually a piece of plate glass slightly larger than the desired membrane and fitted with metal runners 0.025 cm thick along two opposite edges. A metal straightedge or spreading blade is also prepared to ride on the runners. The casting plate and spreading blade are then precooled to a temperature between - 8° and - 11 °C.

3. Place a sufficient quantity of casting solution on the glass plate and wipe the spreading blade along the runners to produce a membrane of uniform thickness. This operation should be carried out in the cold box so that the temperature does not rise above -8 °C.

4. After casting as in step 3, let the membranes on the casting plate remain exposed to the air for 3 or 4 min. Exposures longer than this will reduce both the water flux rate of the membrane and the salt reduction properties.

5. Immerse the membrane-plate assembly in ice water for a period of 30 min to 1 hr.

6. Remove the membrane from the plate and immerse it in hot water for 15 min. The temperature of the water should be held constant between relatively small limits, the exact temperature being

determined by the salinity of the water to be desalted. For seawater desalination, this curing temperature must be in the range 79° to 82°C.

7. When the heating period has been terminated, store the membrane in water at room temperature until used.

In commenting upon the above steps, Loeb and Sourirajan observed that part of the acetone evaporated during step 4, that the magnesium perchlorate was leached out during step 5, and that the final heating temperature of step 6 determined both the membrane's permeability to water and its salt rejection capability for the given salinity. The finished membranes were about 0.004 in. thick and were found to contain none of the magnesium perchlorate from the casting solution. They also reported that these membranes were unidirectional, in that they were relatively impervious to the unwanted ions in saline water only when the air-dried side of the membrane was in contact with the water to be desalted. When reversed, with the casting-plate side of the membrane facing the saline water, the ions would pass through the membrane with relative ease. This unidirectional property was explained by Riley et al. [10] on the basis of studies of membranes cast as noted above and then subjected to electron microscope examination. Their conclusions indicated that the air-dried side of the membranes consisted of a very dense skin of cellulose acetate about 0.25 μm thick, with the balance of the 100 μm thickness consisting of a much more porous and less dense layer of cellulose acetate. From this data it was hypothesized that the exclusion of ions was due to the dense skin; hence the designation of these membranes as "skinned" membranes. In attempting to explain the transmissivity of the membrane, Loeb [1] noted that the magnesium perchlorate was soluble in both the water and the acetone of the casting solution with the result that it was uniformly distributed throughout the cast membrane. When the magnesium perchlorate was leached out in the warm water, therefore, the resulting "holes" were found to be uniformly distributed throughout the membrane.

Loeb attempted to optimize the formulation of the casting solution and the membrane processing steps by varying the proportions of the ingredients and the temperature conditions. He also used various materials other than magnesium perchlorate which were soluble in both water and acetone, and found several formulas which would increase the porosity of the cellulose acetate membrane, but none of these was as satisfactory as magnesium perchlorate from the viewpoints of both transmissivity of water and salt exclusion. Loeb's [1] studies noted that cellulose acetate-butyrate could be substituted for cellulose acetate, and that membranes made with this material did not require the heating of step 5 in the procedure.

Later Majikian, Loeb and McCutchan [11] developed cast membranes using various nonelectrolytes as additives, and of these formamide was found to be the most effective. Successful membranes were cast using many different proportions of the three following ingredients, the limits of composition being: cellulose acetate — 20 to 30%; acetone — 35 to 65%; and formamide — 10 to 40%. The optimum mixture of this type proved to be 25% cellulose acetate, 45% acetone, and 30% formamide, all in percentages by weight. Casting of these membranes was carried out at 23 °C (slightly above room temperature), and was followed by heating which served to increase both the salt exclusion and transmissivity. This casting solution has been used by Johnson et al. [12] in extensive field tests.

As already indicated there are a number of materials besides cellulose acetate which can be made into films possessing osmotic properties, and which have been used successfully in reverse osmosis devices. The DuPont Company has used nylon of the hollow fiber type in desalters, compensating for the very low flux rates characteristic of this material by using very large numbers of small thin-walled tubes. By changing manufacturing variables they were able to vary the water permeability of the nylon tubes to a flux rate between about 0.07 and 0.25 gpd/ft^2. The corresponding amounts of solute not excluded by the membranes ranged between less than 6 to over 30%.

Another unique development in the production of reverse osmosis membranes is the "dynamically formed" membrane conceived at the Oak Ridge National Laboratory. This idea is described by Kraus et al. [13] as being filter aids which are added to the saline water feed stream and are deposited on a coarsely porous structure to form a base on which other additives, such as organic polyelectrolytes, will deposit and form membranes. Although these dynamically formed membranes have not as yet appeared in a production plant, their demonstration illustrates the wide diversity of materials and production techniques that are possible in fabricating reverse osmosis membranes.

B. Explanations of Salt Exclusion

The exact means by which membranes exclude salt has not been agreed upon by researchers in this field, although there are several possible mechanisms. The most obvious concept is that of a simple sieve, in which the holes are intermediate in size between the dissolved ions and the water molecules. For materials such as sodium chloride, which produces ions only slightly larger than the water molecules, the sieve concept would seem very questionable, since ions could pass through almost as freely as water.

A second concept is that of transmission by distillation, that is, by evaporation of water at one side of the membrane, transmission through the membrane as a vapor, and condensation of the vapor at the other side of the membrane. This is not very attractive for explaining water transmission through a polymeric membrane, although Hassler [14] has built a device for distilling water across a thin air gap, the latter serving as the "membrane."

Another concept is that the membrane material is quite wettable, so that water tends to cling to it as an absorbed film. This film then obstructs the "pores" in the membrane and thereby prevents the ions from entering. Reid [8] proposed this "mechanism" and suggests that the water progresses through the membrane by passing from one wetted site to another within the membrane structure. This concept would seem to offer possible explanation for most of the facts relating to water and ion flow through membranes.

Another similar concept is that of the ion-exchange membrane, such as used in electrodialysis. The charged sites in this type of membrane will attract ions of the opposite sign into the membrane, but will not permit them to leave the membrane without an electrical potential. Water can pass through due to a pressure difference across the membrane, thus accounting for both water flow and ion exclusion. While this would be an adequate explanation for ion exclusion characteristics, most reverse osmosis membranes are produced without charged sites. Electrodialysis membranes would not be suitable for reverse osmosis equipment for they are actually constructed to reduce osmotic transfer to the very smallest value. In spite of the lack of a completely acceptable explanation for the osmotic phenomena within the membranes, the fact remains that the membranes used in reverse osmosis do exclude ions and permit water to pass.

C. Membrane Performance

As has been indicated, the two critical operational characteristics of reverse osmosis membranes are their selectivity (exclusion of solute ions) and transmissivity (usually expressed as the water flux rate). These are also most critical from the viewpoint of desalting costs. If a plant is designed to recover approximately half of the saline water as freshwater, the salinity of the brine near the outlet end of the membrane will be approximately twice that at the inlet end. Therefore, in order for the product water to have a specific purity, e.g., less than 500 ppm, the membrane must exclude a large enough fraction of the ions throughout the course of the increasing concentration of the brine stream to produce freshwater of the desired purity. For example, if the brine salinity increases during passage through the device to 5000 ppm at outlet, and the membrane is capable of excluding 90% of the ions, the concentration of the product water will increase from a small value at inlet to 500 ppm at outlet, with the average salinity of the product being appreciably less than 500 ppm. Cellulose acetate membranes of the Loeb-Sourirajan type have been produced which exclude up to 95% of the ions. The formamide type of membrane has a similar ion-exclusion capability. Thus, the necessary ion-exclusion capabilities of available membranes appear adequate for use in once-through desalting devices when applied to moderately brackish water. Seawater desalters would require at least two of the present membranes in series to accomplish the required reduction in salt concentration. It should be noted, however, that there is no theoretical limitation on the degree of salt exclusion possible in a membrane. Researchers are presently pursuing membrane developments which are focused on producing membranes capable of desalting seawater on a once-through basis.

In attempting to rationalize the empirical relationships between membrane exclusion properties and transmissivity for water, algebraic expressions for the flow of water and ions through membranes have been formulated by various authors. Excellent summaries are given by Merten [3] and Lonsdale [15]. The simplest equations are based on the assumption that the flows of water and dissolved ions are independent of each other and that they take place by diffusion through the membrane. The equations resulting from this reasoning have the general forms of the following:

$$F_1 = -K_1(\Delta P - \Delta \pi) \tag{5}$$

= flow of solvent (water)

$$F_2 = -K_2(\Delta C) \tag{6}$$

= flow of solute (ions)

where K_1 and K_2 are constants, ΔP is pressure difference across membrane, $\Delta\pi$ is difference in osmotic pressures of solutions on the two sides of the membrane, and ΔC is difference in concentration of dissolved salts on the two sides of the membrane. The negative algebraic sign indicates that the flow in reverse osmosis takes place in the direction of decreasing pressures and concentrations. The constancy of K_1 and K_2 presume that the membrane properties are independent of pressure and concentration.

However, the flow rates of water and dissolved materials are not truly independent of each other as assumed above, so considerable study has been directed toward applying irreversible thermodynamic reasoning to the relationship, using coupled flows. Lonsdale [15] has investigated this application and summarizes his reasoning by equations for solvent and solute flow which have the following forms:

$$F_3 = -K_3(\Delta P - \sigma\Delta\pi) \tag{7}$$

= volumetric flow of solvent through the membrane

$$F_4 = K_4(1 - \sigma)F_3 - K_5(\Delta C) \tag{8}$$

= flow of solute

where σ measures the coupling of solvent to the solute flows within the membrane and K_3, K_4, and K_5 are constants. It will be noted that when there is no coupling of flows, i.e., $\sigma = 1$, these equations are nearly identical with the two previous equations.

An examination of the equations indicates that the flow of water is proportional to the pressure difference and that the flow of solute is nearly independent of pressure but proportional to the concentration difference. The presence of the first term on the right-hand side of Eq. (8) indicates that the flow of ions through the membrane increases as the solvent water increases, thus increasing the salinity of the product.

The above theoretical conclusion seems to be borne out in practice, since experimentation indicates an inverse relationship between salt exclusion and water flow; i.e., as the water flux for a given pressure increases, the percentage of salts excluded decreases. For example, Manjikian et al. [11] reported that changing the formamide proportion in the casting solution from 25 to 30% doubled the water flux but decreased the salt excluded from 94.7 to 91.8%. Thus the theoretical conclusions drawn above are borne out and the need for precise formulation and processing of reverse osmosis membranes is emphasized.

Although the operational characteristics of ion exclusion and flux rate determine the applicability of a given membrane to a particular desalination situation, the useful life of a membrane, as determined by its salt rejection over a long period of time, is of great importance from an economic viewpoint. Some of the early membranes failed in a few months' time, and much study has been devoted to determining the causes of failure. One conclusion has been that the pH of the water is a contributing factor, with longer service being observed when the pH is low. For example, Johnson et al. [12] tested tubular assemblies with seawater and found that a pH of 7.8 resulted in a definite increase in the salinity of the product water in two weeks' service, whereas a pH of 4.5 resulted in no change for that same period of time. It has been argued that the deterioration of the membrane is due to hydrolysis which is accelerated by high values of pH. Experience with the tubular membrane assembly installed at Coalinga, California, where the community was supplied with freshwater from their brackish sources for a number of years, has been reported by Johnson et al. [12]. During a period of three and one-half years they observed that a few of the membrane tubes remained operable for as long as two years, with the average life of all the tubes being over 300 days. The source of the feed was well water with an average pH of 7.6 and a salinity of 2450 ppm. Their report seems to indicate that satisfactory performance of membranes of the cellulose acetate type modified with formamide can be expected to last about one year.

D. Polarization

The performance of a reverse osmosis membrane is affected not only by the characteristics of the membrane itself but also by the conditions of flow of the saline water across the membrane surface, under certain conditions of which polarization may occur. "Polarization" in reverse osmosis devices is the decrease or cessation of water flux through the membrane due to the local increase of saline water concentration at the surface of the membrane. This local increase in concentration is inevitable, since passage of water through the membrane leaves behind the ions which were dissolved in it. These ions must be moved away from the membrane surface rapidly for maximum water flux through the membrane. Their rapid dispersal is complicated by the fact that the turbulent flow of the saline water stream is bounded by a thin viscous boundary layer next to the membrane surface, and through this layer ions move by diffusion rather than by the eddying motion of the main stream. Movement by diffusion is brought about by differences in concentration between the membrane surface and the turbulent core, and is slow as compared to the motion by turbulence. Increased velocity reduces the thickness of the boundary layer and hence the distance through which the ions have to move by diffusion. For this reason designers attempt to use velocities as high as possible in order to minimize the thickness of the boundary layer, and to disturb the inertia of the boundary layer by inserting turbulence promoters in the flow channel.

Besides the increase in average salinity as water progresses through the membrane-lined passages there is thus a further increase due to the concentration gradient at the membrane surface. These increased concentrations result in increases in osmotic pressure, thus decreasing the pressure differential available for the flow of water through the membrane. Should this pressure differential be reduced to zero the flow would cease and polarization would be complete. However, in reverse osmosis design there is always some remaining pressure differential, so actually polarization will be only partial, but results in producing an accompanying water flux which will be less than that which would result from merely the increased average concentration of the brine. It is clear that any desalting device must take into account the average increase in concentration and also make some allowance for the concentration gradient which produces the polarization effect noted above.

IV. REVERSE OSMOSIS DEVICES USED FOR DESALINATION

Devices suitable for the desalination of water by reverse osmosis involve the arrangement of membranes and their supporting surfaces so that saline water under high pressure can be channeled across the membrane and the freshwater can be collected without contamination by the brine. The high pressures required for operation of the device, combined with the relatively fragile nature of the membranes, require that there be strong supports for the membranes. This requirement has led many of the developers of this process to prefer the tubular form of the membrane, so that the comparatively fragile membrane can be contained within a strong supporting tube with the high pressure inside of the tubular membrane. Other developers have chosen to use flat membranes, reasoning that these membranes, being more readily adapted to quantity production methods, would be less expensive to manufacture than would tubular membranes. Devices using flat membranes include either plate supports or tubular supports, such as the spiral-wound type.

A. Units Employing Flat Membranes

1. Plate-and-Frame Assemblies

The first experimental unit constructed by Loeb et al. [16] was designed to use the concept of the plate-and-frame filter press. In this design, shown schematically in Fig. 14-2, a membrane assembly consists of a circular metal support plate, with a membrane on each side of the plate and several layers of nylon parchment between the plate and the membrane to collect the freshwater as it seeps through the membrane. A reverse osmosis stack contains several of these assemblies, the membranes of which are sealed against their support plate at their periphery and at either a central hole or a hole near the periphery with O rings, so that saline water on the saline side of the membrane cannot mix with the product water next to the plate. The filtered water, collected by the porous nylon parchment, is conducted through several narrow slits cut into the surface of the support plate to central radial holes drilled from the edge of the plate, and through these holes flows to the product water collection center. The saline water flows through the assembly stack in the space between the adjacent membranes, moving from one inter-membrane space to the next through the alternately placed assembly holes mentioned above. The assemblies are separated by spiral inserts between the adjacent membranes which cause the water to flow in a spiral path, these spirals alternating, one to a central hole and the other to a peripheral hole, thus promoting a circuitous saline water path over the face of the membrane en route to the exit of the stack. A unit of this design was operated successfully for a number of years, some results of which are discussed in Ref. [16].

Another configuration of the plate-and-frame design was developed by the Aerojet-General Corporation [17], which constructed and operated experimental units as large as 50, 000 gpd. Their 50, 000 gpd unit contained in all over 200 membrane assemblies consisting of 36 in. diameter plastic plates supporting on each face a reverse osmosis membrane of the same outer dimension and an inner diameter of 9 in., and producing a total active desalting area of about 2500 ft^2. Through a center hole in each plate ran a 3.3 in. diameter tube through which the product water flowed to a collection tank. The membrane assemblies were grouped in nine chambers under pressure, and the saline water in each chamber flowed in parallel between the membrane assemblies where it seeped through the membranes and a porous layer to interior channels and radial holes in the plates, and, through these small holes, to the central collection tube. In order to

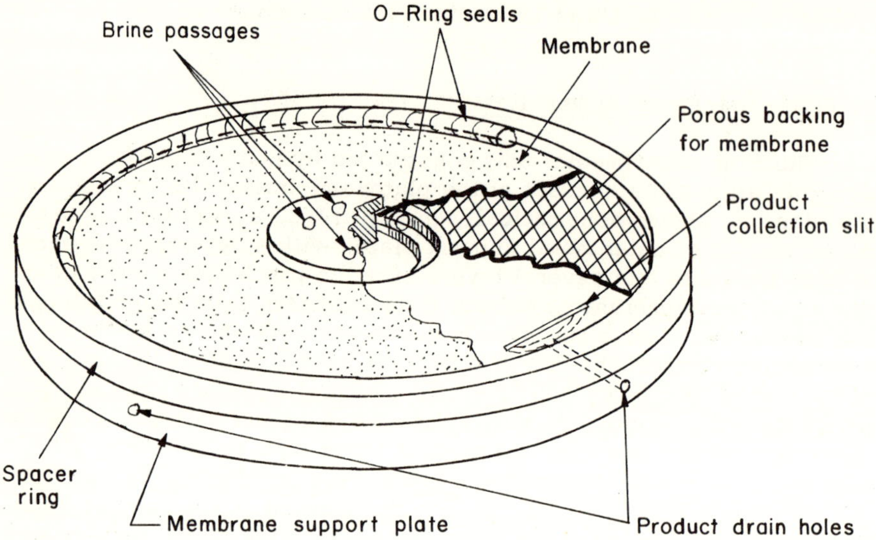

FIG. 14-2. Membrane and support plate for plate-and-frame type of reverse osmosis desalter (schematic only). Note that figure shows top of plate and membrane assembly only. Membrane on bottom of plate is not shown, nor is spiral insert for guiding the saline water across the membrane surfaces. The center plug containing the brine passages is mated with a similar plug on the lower side of the plate, the two plugs being drawn together to compress the small O rings and thus seal the inner edges of the membranes.

ensure good circulation, the saline water was caused to flow in parallel spiral paths across the faces of all the membranes in a single chamber, alternating in direction in each successive chamber from the center to the outer periphery and vice versa. Spacing between membrane plates was provided by rims around their inner and outer peripheries, gaps in the outer rim permitting the saline water to pass between the inter-membrane spaces, and the annular space between the outer edges of the plates and the walls of the pressure chamber. Leakage of the saline water into the product stream was prevented by bonding the ring-shaped membranes to the plastic support plates along the inner and outer peripheries and placing an O-ring seal between the inner rim and the ring of small holes through the plates which allowed the saline water to be distributed to or collected from all of the plates in a given chamber. For the purpose of testing the membranes as well as the effectiveness of the process with several different types of water, this experimental plant was mounted on a semi-trailer, together with filters, pumps, and other equipment, and operated at a number of different locations. Details of construction will be found in Ref. [17] and the results of tests in Ref. [18].

2. Spiral-Wound Blanket Assemblies

The problems of edge sealing and freshwater collection inherent in the plate-and-frame assemblies noted above led the researchers of the Gulf General Atomic Company [19], to develop the concept of the spiral-wound blanket as a means of utilizing flat membranes. As shown in Fig. 14-3, these units consist of a central pipe for collecting the freshwater and two membranes separated by a porous layer for receiving and conducting the freshwater to holes in the pipe. The two membranes, with the saline water side outermost, are sealed together around the three edges not connected to the pipe so that the saline water under pressure will be in contact with their outer surfaces. The membrane blanket, sealed to the pipe along the inner edge, is then wrapped around the pipe, with a screen on one side to separate the several layers of the spiral and provide a passage for the saline water across the membrane surfaces. This entire assembly is then inserted inside of a pipe just large enough to contain it, with the ends of the freshwater pipe extending through the end seals of the larger pipe. In operation the saline water is pumped into one end of the large pipe and exits from the other end. The fit of the spiral blanket within the large pipe is sufficiently snug that the saline water must flow through the spaces between the several turns of the membrane blanket, thus bringing it into contact with the membrane surfaces and permitting the reverse osmosis flow of freshwater to occur. This freshwater is conducted to the central pipe by the porous material contained within the blanket. The actual units manufactured by the above company consist of several of the basic units enclosed in a single large pipe, with the freshwater collection pipes connected in series. In such assemblies the first spiral-wound unit is exposed to saline water of the initial salinity and the successive units to successively higher salinities. Several successful types of units of this design have been built in various sizes.

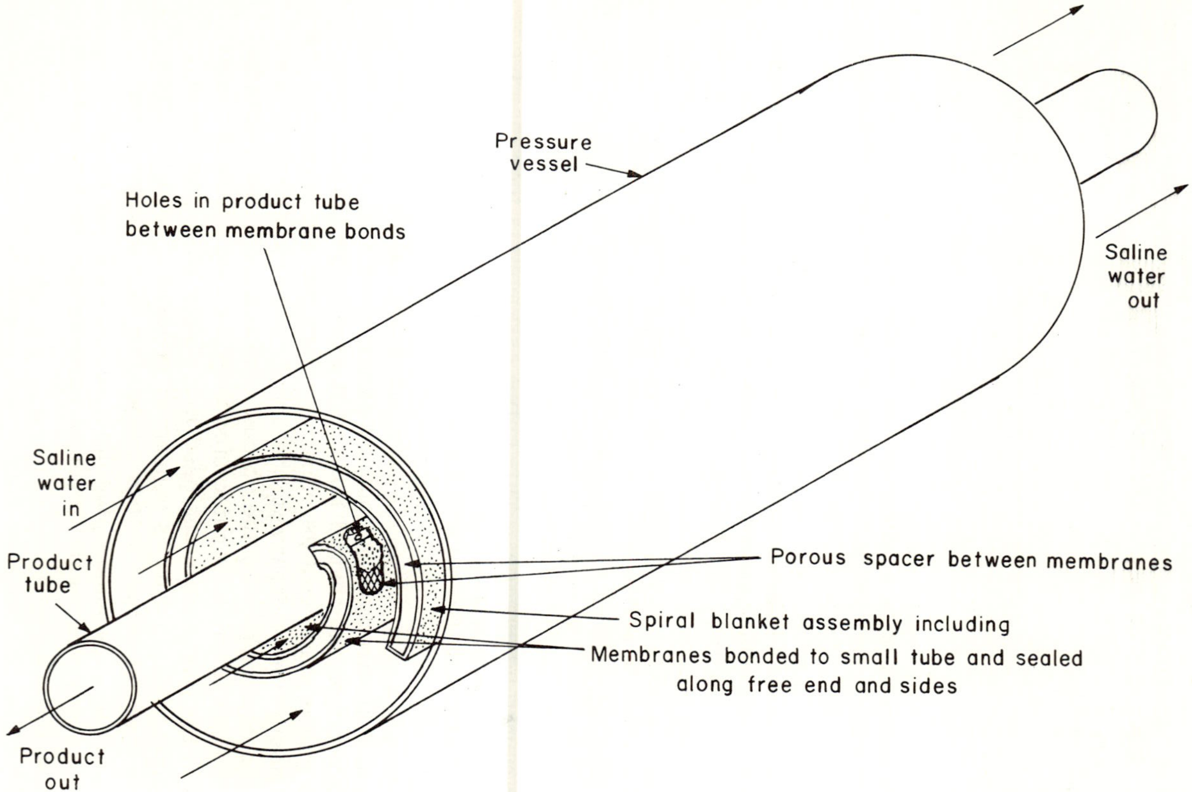

FIG. 14-3. Spiral blanket type of reverse osmosis desalter (schematic only). Note that the end bell of the pressure vessel has been omitted in order to show the blanket arrangement. Also omitted is the mesh spacer used to separate one turn of the spiral from the next and to provide flow channels between turns for the saline water. While the figure shows only one turn of the spiral for clarity, the production units use many turns in order to provide as much membrane area as possible for a given volume of pressure vessel.

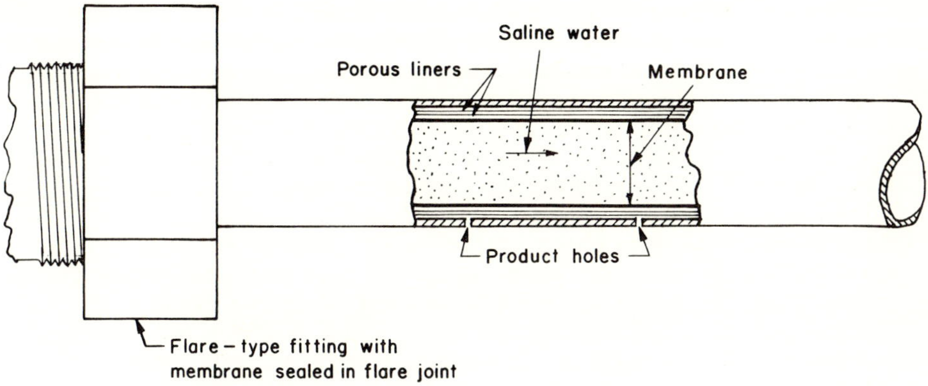

FIG. 14-4. Tubular type reverse osmosis desalter (schematic only). Note that the sheath for collecting the product water from the holes in the copper support tube is not shown.

B. Units Employing Tubular Membranes

1. Glass Fiber Tubes with Membranes Cast "In Place"

This configuration was conceived by Havens [20] and has been used in several commercial plants. In this design porous glass fiber tubes, $\frac{1}{2}$ in. in diameter, are filament-wound from fine glass fibers and then bonded with resin, resulting in an inert tube which has considerable strength and is sufficiently porous to permit the product water to pass through it radially. After fabrication of the tube a modified cellulose acetate lining is cast inside of it and processed to be a reverse osmosis membrane, with the selective side forming the central saline water passage. Saline water pumped along the inside of the tube under sufficient pressure will then cause freshwater to pass through the tube radially so that it can be collected from the outside of the tube. Where a number of tubes are used they can be arranged in series by connecting them end to end, or in parallel by placing them between headers, as in a shell-and-tube heat exchanger. The major problem in either case is the sealing of the ends of the membrane linings to prevent contamination of the freshwater by leakage of the saline water. This has been accomplished by the designers for both of the above configurations.

2. Tubular Membranes Inserted in Metal Tubes

There are several reverse osmosis desalting units in which precast tubular membranes are inserted into metal tubes, one example being the units installed at Coalinga, California, by Loeb [21]. The desalting portion of this plant consisted of 112 copper tubes 1 in. O.D. by 10 ft long, in which the interior of each tube is lined with a tubular membrane of modified cellulose acetate. The general arrangement of these tubes is shown in Fig. 14-4. In the assembly of this plant the membranes were cast, soaked in cold water for at least 1 hr, wrapped tightly in two or more thicknesses of porous nylon cloth, and slipped into the above-mentioned metal tubes in which small holes had been drilled through the tube walls every 3 in. throughout their length, and finally heated by pumping warm water through the tubes. When pressurized saline water is pumped through the tube, desalted water, passing through the membrane tube, flows through the porous nylon cloth to the nearest of the small holes, and drips through these holes to be collected by a plastic sheath which encloses the copper tube and channels the product water into a collection trough at the end of the assembly. In the Coalinga plant the tubes were mounted horizontally in two banks, one above the other. The tubes in each bank were connected in series by U bends, and the joints sealed against leakage by flared fittings so arranged that the membrane tube was clamped in the "flared" portion. If this had been designed as a production plant rather than an experimental one, the plastic sheath probably would have been replaced by a metal housing. Other plants using inserted membrane tubes have used tube sheets, shells, and water boxes, assembled in much the same manner as tubular heat exchangers.

3. Self-Supporting Tubular Membranes

When membrane tubes are made small enough in diameter, they can be strong enough to withstand considerable pressure. This idea was first applied to desalination by Mahon [22] who pointed out that capillary tubes of cellulose triacetate, 70 to 100 μm O.D. with walls 5 to 10 μm thick would serve to desalt water at a high rate per unit volume of apparatus even though the flux per unit area was small. More recently the DuPont Company has marketed a device in which capillary fibers of nylon are used to desalt water. Cooke [23] describes these units as containing fibers 45 μm O.D. and 24 μm I.D. Large numbers of these fibers are bundled together, the bundle doubled so that all tube ends are together. These ends are sealed with an epoxy resin, forming a tube-end block, the exterior surface of which is cut away to open the tube ends. The bundle is enclosed in a shell, similar to a U-tube heat exchanger, into which saline water is pumped under pressure, infiltrating the shell side of the tubes. Freshwater seeps through the tube walls, flows within the capillary tubes and out through their open ends to be collected as product. A unit of this type, 14 in. in diameter and 10 ft long, is described as containing 28×10^6 fibers with a total area of 84, 500 ft^2 of membrane surface, and is rated as capable of producing 8600 gpd of desalted water from a supply of 1500 ppm salinity using pressures of 600 psi.

The previous descriptions of reverse osmosis desalters are representative of the several configurations which have been used in actual plants. The patents covering these and many other desalting devices have been summarized by McDermott [24].

V. DESIGN CONSIDERATIONS

The designer of a reverse osmosis desalting plant is generally confronted with the task of determining the membrane area required for a given production rate of freshwater, and of sizing the saline water passages in the device to achieve this rate of production. The membrane area needed will be inversely proportional to the product flux of the membrane, i.e., the rate of flow of product water through unit area of membrane. While each manufacturer will have available precise empirical information concerning these values, it is possible to indicate approximate ranges of the quantities involved. For flat membranes and tubular membranes of $\frac{1}{2}$ in. and larger diameters, the membrane product flux will vary between 10 and 50 gpd/ft^2 for saline water desalting, and between 1.5 and 6 gpd/ft^2 for seawater desalting. The product flux rates for capillary membranes of nylon will be much less, being smaller than 0.1 gpd/ft^2 for saline water.

These variations in flux rates are dependent upon the membrane composition, the curing temperature of the membrane, the temperature of the saline water being processed, and the pressure applied to the desalting unit. Composition and curing temperatures, as previously explained, affect the desired selectivity properties of a membrane for a particular water supply, so that knowledge of the water supply is necessary for the preparation of the membrane. Estimates of the effect of water supply temperature upon flux rates can be based on the observation that a change of 1 °F in temperature will change the product flux rate by 1.5 to 3%. The pressure to be applied in a plant is arbitrarily selected well above the osmotic pressure level and is usually approximately 600 psi for most brackish water installations. The effect of pressure on product flux can be approximated by the relation:

$$\text{Product flux rate} = K(P - \pi)$$

in which K is the membrane constant, P is the applied pressure, and π is the osmotic pressure of saline water. This relation is applicable strictly to ideal, perfectly selective membranes, so that the behavior of actual membranes in which selectivity is less than perfect will deviate from it. Nevertheless, this equation enables the designer to estimate the change in product flux to be achieved when pressure is changed from some value for which the product flux is known. Any change in pressure also changes the ion exclusion slightly, the latter quantity increasing with an increase in pressure.

As has been described, polarization in a reverse osmosis installation reduces its effectiveness, so the design of a desalting unit requires the proportioning of the saline water passages such that polarization may be limited to as small a value as possible. To accomplish this, the degree of turbulence in the flowing saline water should be high in order to limit the thickness of the boundary layer and hence its tendency toward polarization. In flat plate units this is achieved by turbulence promoters incorporated into the spacer devices. In tubular units turbulence can be ensured when water velocities are sufficiently large that the corresponding dimensionless Reynolds' numbers are considerably in excess of 2500.

VI. ADDITIONAL COMPONENTS OF REVERSE OSMOSIS DESALTING PLANTS

A plant using membrane units as described in the previous sections also includes several additional components, a representative layout of which is shown in Fig. 14-5. In this diagram it will be seen that there is a section for the pretreatment of the saline water, and also a high pressure pump and a back-pressure regulator valve for maintaining pressure on the saline water in the membrane units. There have been several suggestions to replace the back-pressure regulating valve by a hydraulic turbine so as to recover some of the pumping work as the waste water is being expanded to atmospheric pressure. Although this proposal has apparent merit it has not appeared in any installation so far, probably because of the relatively

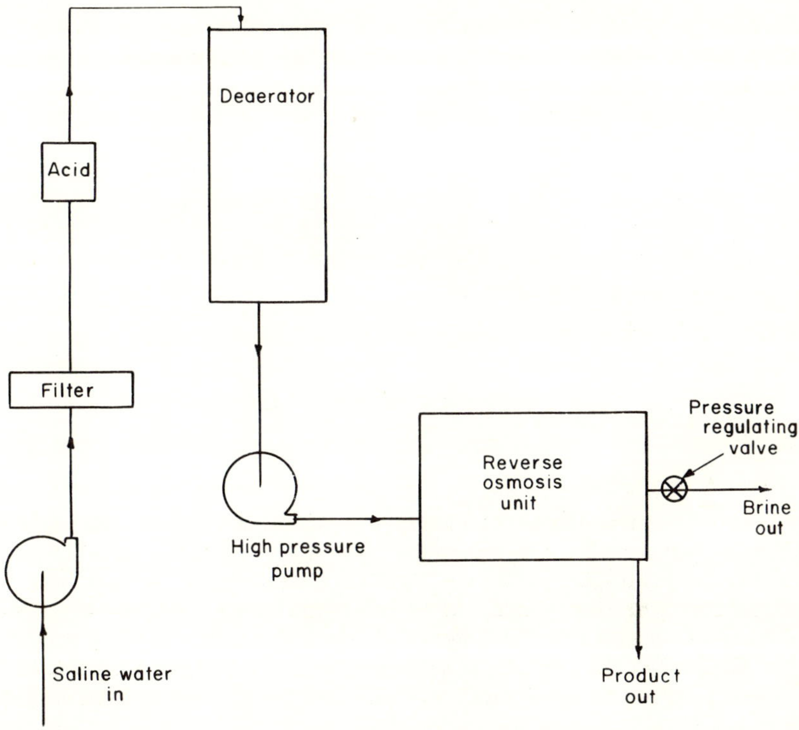

FIG. 14-5. Plant arrangement for reverse osmosis desalting of water (schematic only).

small size of these installations and the corrosive nature of the waste brine.

Pretreatment of the incoming saline water is designed to prevent clogging of the passages in the membrane unit, to control the hydrolysis of the membranes, and to minimize the possible corrosion of the pump and other metallic portions of the saline water flow path. To prevent clogging of the passages the incoming saline water must be filtered to remove any suspended matter. Filters, sized to remove particles of 10 μm diameter, have been found satisfactory for most units, with 5-μm filters specified for the capillary devices. Cartridge filters are used for small units, and diatomaceous earth or sand filters for larger plants. Besides the suspended matter, some waters contain scale-forming chemicals. Precipitation of these chemicals can be prevented by the same treatments as described for distillers, as for example, the addition of polyphosphate compounds to water containing calcium and carbonate ions. Since hydrolysis of the membranes is related to the alkalinity of the saline water, it is usual to adjust the pH of the water to a value slightly on the acid side of the neutral point. Acid is added in quantities sufficient to lower the pH to the range of 4.5 to 6.5. Corrosion of the metallic portions of the flow path is accelerated by the presence of oxygen in the saline water. For this reason means for removing oxygen must be included in the pretreatment system, such as the deaerator shown in Fig. 14-5, although such deaerators have seldom been included in the design of the small plants built in the past.

VII. COST OF DESALTED WATER

Only a few reverse osmosis desalting plants have been in operation long enough to have satisfactory cost records, and even their figures are colored by the experimental nature of the plants. The costs of the 5000-gpd plant at Coalinga, California, which was operated for over five years, have been analyzed by Johnson et al. [12] after three and one-half years of its operation. From their study it was estimated that the cost of water from that plant was $1.94/kgal, but should probably have been $1.33/kgal for the same size and design of a plant intended for water production rather than for research and development. Both cost figures included $0.175/kgal for data taking labor. If this amount is deducted from the above production plant cost, the total cost estimate can be summarized as follows:

Amortization, taxes, interest	$0.430
Membrane replacement	0.266
Maintenance	0.298
Operation — power and filters	0.163
Total	$1.157

This high figure is due primarily to the very small size of the plant. For a production plant of ten times the capacity of the experimental unit, but of similar design, Johnson et al. estimated a water cost as low as $0.50/kgal. This latter figure approximates that given by Lacey [2] for large production plants using various types of reverse osmosis units. The extensive research and development efforts presently being expended should serve to reduce these figures in the future.

REFERENCES

1. S. Loeb and S. Sourirajan, Sea Water Demineralization by Means of a Semipermeable Membrane, UCLA Engineering Rept. No. 60-60, 1960, University of California, Los Angeles, California.
2. R. E. Lacey and S. Loeb, Industrial Processing with Membranes, Wiley-Interscience, New York, 1972.
3. U. Merten, Desalination by Reverse Osmosis, The MIT Press, Cambridge, Massachusetts, 1966.
4. A. Findlay, Osmotic Pressure, Longmans Green, London, 1913.
5. J. H. van't Hoff, Z. Phys. Chem., 1, 1887, p. 481.
6. P. Debye and E. Hückel, Phys. Z., 24, 1923, p. 185.
7. R. A. Robinson and R. H. Stokes, Electrolyte Solutions, Academic Press, New York, 1959.
8. C. E. Reid et al., Water and Ion Flow Through Imperfect Osmotic Membranes, Res. Develop. Progr. Rept., No. 16, Office of Saline Water, U. S. Dept. Interior.
9. E. B. Millard, Physical Chemistry for Colleges, 1926, p. 161.
10. R. L. Riley et al., Replication Electron Microscopy of Cellulose Acetate Osmotic Membranes, Desalination, 1 (No. 1), 1966, pp. 30-34.
11. S. Manjikian, S. Loeb, and J. W. McCutchan, Improvements in Fabrication Techniques for Reverse Osmosis Desalination Membranes, Proc. 1st Intern. Symp. Water Desalination, Washington, D. C., 2, 1959, p. 159.
12. J. S. Johnson, J. W. McCutchan, and D. N. Bennion, $3\frac{1}{2}$ Years' Experience with Reverse Osmosis at Coalinga, California, School of Engineering and Applied Science Rept. No. 69-45, July, 1969, Univ. of California, Los Angeles, California.
13. K. A. Kraus et al., Hyperfiltration Studies X: Hyperfiltration with Dynamically Formed Membranes, Desalination, 2 (No. 2), 1967, pp. 243-266.
14. G. Hassler and J. W. McCutchan, Osmosis Through a Vapor Gap Supported by Capillarity, Advan. Chem. Ser., 27, 1960, pp. 192-205.
15. H. K. Lonsdale, Theory and Practise of Reverse Osmosis and Ultrafiltration, Chap. 8, Industrial Processing with Membranes (R. E. Lacey and S. Loeb, Eds.), 1972.
16. S. Loeb and F. Milstein, Design, Development and Testing of a 500 Gallon per Day Osmotic Desalination Cell, Dept. of Engineering Rept. No. 62-52, 1962, Univ. of California, Los Angeles, California.

17. E.R. Watson, Design and Fabrication of a 50,000 gpd Portable Reverse Osmosis Pilot Plant, Res. Develop. Progr. Rept., No. 431, May 1969, Office of Saline Water, U.S. Dept. Interior.
18. Aerojet-General Corporation, Operation of a 50,000 gpd Portable Reverse Osmosis Pilot Plant, Res. Develop. Progr. Rept., No. 430, 1969, Office of Saline Water, U.S. Dept. Interior.
19. T.J. Larson, Reverse Osmosis Pilot Plant Operation: A Spiral Model Concept, Desalination, 7 (No. 2), 1970, pp. 187-199.
20. G.G. Havens, Desalination by Reverse Osmosis (V. Merten, Ed.), 1966, Chap. 7.
21. S. Loeb, A Composite Tubular Assembly for Reverse Osmosis Desalination, Desalination, 1 (No. 1), 1966, pp. 35-49.
22. H.I. Mahon, Hollow Fibers as Membranes for Reverse Osmosis, Nat. Acad. of Science, Nat. Research Council, Publ. No. 942, 1961, pp. 345-354.
23. W.P. Cooke, Hollow Fiber Permeators in Industrial Waste Stream Separations, Desalination, 7 (No. 1), 1969, pp. 31-46.
24. J. McDermott, Desalination by Reverse Osmosis, Noyes Data Corp., 1970, 20 pp.

AUTHOR INDEX

Numbers in parentheses are reference numbers and indicate that an author's work is referred to although his name is not cited in the text. Underlined numbers give the page on which the complete reference is listed.

SUBJECT INDEX